Recording and Voice Processing 1

# Recording and Voice Processing 1

*History and Generalities*

Jean-Michel Réveillac

iSTE

WILEY

First published 2021 in Great Britain and the United States by ISTE Ltd and John Wiley & Sons, Inc.



ISTE Ltd
27-37 St George's Road
London SW19 4EU
UK

www.iste.co.uk

John Wiley & Sons, Inc.
111 River Street
Hoboken, NJ 07030
USA

www.wiley.com



Library of Congress Control Number: 2021942968

British Library Cataloguing-in-Publication Data
A CIP record for this book is available from the British Library
ISBN 978-1-78630-670-8

MIX
Paper from responsible sources
FSC® C013604

# Contents

# Preface

If you want to know if this book is for you, how it is constructed and organized, what it contains and what conventions will be used, you've come to the right place, this is the place to start.

## Target audience

This book is intended for all those who, amateur or professional, are interested in sound recording, recording and mixing in the field of singing and voice or musicians, performers, commentators and composers.

The work presented in some sections requires minimum knowledge in the field of acoustics and digital audio.

You must have a good knowledge of your computer's operating system (paths, folders and directories, files, names, extensions, copying, moving, etc.) and know how to handle a DAW (Digital Audio Workstation), such as Avid Pro Tools, Apple Logic Pro X, Ableton Live, Steinberg Cubase, FL Studio, MOTU Digital performer, Cockos Reaper, etc., or a digital integrated studio, such as Tascam DP-03SD, Tascam DP32, Roland VS-1680, Akai DPS16HD, Yamaha AW4416, etc.

## Structure and contents of the book

This work is composed of two volumes:

1) history and generalities;

2) studio work.

Volume 1 presents a preface, specifying the contents and the writing conventions used, then an introduction followed by four chapters, a conclusion and five appendices:

– recording history;

– voice;

– microphones;

– acoustic environment.

The conclusion summarizes the main topics discussed and introduces the concepts that will be addressed in the second volume.

Appendices 1–5 provide some additional information. You will find in this order:

– sound unit;

– audio connectivity;

– audio processing plugins;

– tube and JFET mic amplifiers;

– microphone pairs.

Volume 2 presents a preface and an introduction identical to those of Volume 1 followed by four chapters, a conclusion and five appendices:

– processing hardware and software;

– configuration and audio channel;

– voice recording;

– special effects.

Appendices 1–4 are taken from Volume 1 to complement the previous chapters by including:

– sound unit;

– audio connectivity;

– audio processing plugins;

– microphone pairs.

Appendix 5 of Volume 2 provides details on the types of software plugins available from different vendors and operating systems.

The conclusion sheds light on the whole book and gives a brief overview of the future evolution of voice recording.

Each volume can be read separately. While there are concepts that are dependent on another chapter, references to the relevant sections are given. However, the first two chapters of Volume 1, devoted to the history of recording and to the human voice, provide a contextual basis for the understanding of several notions that you will find in the following chapters.

If you're a novice on the subject, I strongly advise you to read them first, to discover the basics of the subject of this book.

For the others, I hope that you will discover new notions that will enrich your knowledge.

At the end of each volume, you will find a reference list and a list of Internet links.

A glossary is also present; it will explain some acronyms and terminologies very specific to sound recording, recording and mixing.

## Conventions

This book uses the following typographical conventions:

*Italics*, which are reserved for important terms used for the first time in the text which may be present in the glossary at the end of the book, mathematical terms, comments, equations, expressions or variables.

Remarks are indicated by the presence of the keyword: NOTE. They complete the explanations already provided.

The figures and tables all have a legend that is often useful for understanding.

## Vocabulary and definition

As with all techniques, voice recording has its own vocabulary. Certain words, acronyms, abbreviations, initialisms and proper names are not always familiar and will be included in the glossary.

## Acknowledgments

I would especially like to thank the ISTE Ltd team and my editor Chantal Ménascé, who trusted me.

Finally, I would like to thank my wife, Vanna, and my daughters Océane and Léa who supported me throughout the writing of this book.

August 2021

# Introduction

Since Plato and Aristotle, we have often heard this quote, "*The voice is the mirror of the soul*". It is true that before understanding the meaning of a word, we must understand its emotional side; we recognize the intonation that makes a voice more or less unique. The paralinguistic characteristics of a voice, that personalize it, take priority over the meaning of the message that is delivered. They express the emotional aspect of the interlocutor and favor the transmission of emotions.

It is also necessary to take the environment into consideration; the voice exists only through its medium, the air which surrounds us. The numerous acoustic variables, vocal cords and different morpho-anatomical cavities of the individual form the timbre of the voice, which can be modified during its diffusion by the environmental medium in which it is established.

Each individual has a different timbre that represents a sound signature. The voice is the reflection of the body and the spirit, but let us not assume that its characteristics are as static and reliable as the genetic code or fingerprints. They are variable, determined by numerous parameters linked to morphology, imagination, environment, affect and even gestures, even if many of these elements do not affect the hard core that builds the essence of a voice.

The recorded voice brings different sensations that often flout the canons of the direct voice, spoken or sung, but it is far from being unpleasant and it has become, by force of things, a standard. Our world is constantly saturated with music or recorded voices; few people can claim to have heard a particular performer with his or her live voice. Voices are optimized for recording and broadcasting; they must remain intelligible and are sometimes enriched, or even placed in a virtual environment, but our ears have become accustomed.

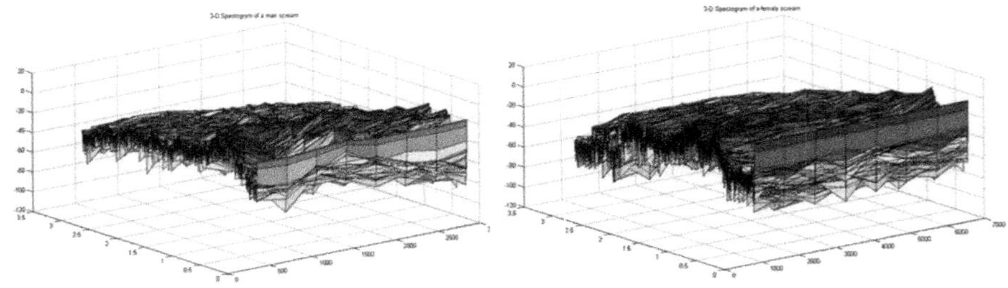

**Figure I.1.** *Vocal differences in the same sentence. 3D spectrogram of a male voice, on the left, and a female voice, on the right (source: www.projectrhea.org). For a color version of this figure, see www.iste.co.uk/reveillac/recording.zip*

It is in part the facts I have just mentioned that led me to write this book, to which I would add the pleasure of bringing, to one or more performers, the possibility of discovering and hearing themselves on a medium such as a CD, DVD, streaming platform, podcast, etc.

I will not forget the listeners nor the public for whom we all work *in fine*: artists, commentators, singers, sound engineers, technicians, etc.

This work, as its title suggests, is above all devoted to the recording and processing of the voice. I have tried to aggregate a set of elements that contribute and intertwine in this same field, taking care of the ins and outs.

I remain convinced that in order to understand the techniques and handle certain hardware or software, it is useful to know what surrounds them. Throughout the book, I have often specified the historical context without forgetting the personages to whom we owe these wonderful inventions. They made it possible for us to listen to the voice, whether spoken, whispered, sung or shouted. At the beginning of the 21st century, and for more than 120 years already, this seems so common, so futile, so banal, but what would our ancestors from the past centuries think of it?

Volume 1 of this book, more oriented towards theory and generalities, will begin with a short history of the recording of the spoken and sung voice from the first machines until today, which will be followed by an outline of the mechanisms and specificities that characterize the human voice.

With these elements, we will approach microphones, an essential and necessary tool for voice capture, to then understand the recording environment, whose influence creates the conditions for a good recording.

**Figure I.2.** *A wire recorder, the Webster Chicago 180-1 (1949). For a color version of this figure, see www.iste.co.uk/reveillac/recording.zip*

**Figure I.3.** *The famous Neumann U67 tube microphone and its power supply. For a color version of this figure, see www.iste.co.uk/reveillac/recording.zip*

Volume 2, oriented towards studio work, will describe the hardware and software tools that are useful for efficient processing during recording and will include them in the audio chain that will support them. Different examples of configurations will be presented, from the most modest to the most advanced.

After this first approach, we will be ready to carry out the sound recording, not without taking some precautions with regard to the performer and their immediate environment. We will place the various essential materials while respecting the rules necessary for a correct recording.

Later, we will discuss the constraints of recording a group of performers, a choir and a voice-over.

Once the sound recording has been done, it is time for processing and mixing by implementing various hardware or software processes that will ensure a correct sound rendering, while respecting the style and coloration you want to apply.

**Figure I.4.** *The famous EMI Abbey Road studio 2 control room with its REDD.37 console (1958–1964)*

Finally, additional effects or corrections can be made to finalize your project.

I hope that the eight chapters that make up these two volumes will provide you with a broader and more precise knowledge of vocal recording, a key area of our daily ultra-mediatized world that seems trivial, but whose mastery is not trivial.

# 1

# Recording History

Before approaching the voice and all the tools that will capture, process, modify, mix and even embellish it, before putting it on a medium that everyone will be able to listen to, I think it is necessary that you discover the eventful history of sound recording from its beginnings until today.

## 1.1. In the beginning was the phonautograph

In 1857, the Frenchman Edouard Léon Scott de Martinville[1], a typographer, invented the "*phonautograph*", a device that records the voice. His machine consists of a membrane with a stylus, placed at the end of an acoustic horn. The stylus records the vibrations received by the membrane on a cylinder coated with black smoke.

Sound was recorded but the question of how to read the recordings still remained; half of the work was thus carried out. E. de Martinville nevertheless filed a patent for this invention on March 25, 1857. He improved his invention the following year by associating with the manufacturer Rudolf Koenig.

The phonautograph received the support of many scientists, but unfortunately not of the financiers with whose money he could have used to market his device. Ruined by his research, which he continued, he died in 1879 forgotten by all.

1 Édouard-Léon Scott de Martinville, 1817–1879, French typographer and inventor who worked in the world of sound.

**Figure 1.1.** *The phonautograph. For a color version of this figure, see www.iste.co.uk/reveillac/recording.zip*

## 1.2. When it really started

On April 18, 1877, the Frenchman Charles Cros[2] deposited, in a sealed envelope, at the *Académie de Sciences*, a process allowing for a voice to be recorded on a cylinder and on a disk that he named the "*paleophone*". The device was named the "*phonograph*" by the abbot Le Blanc (his real surname was Lenoir), who described it in a "*La semaine du Clergé*" article in October of the same year.

**Figure 1.2**. *Charles Cros' "paleophone"*

---

2 Charles Cros, 1842–1888, French poet, scientist and inventor who worked in the field of sound reproduction and color photography.

On December 3, the *Académie des Sciences* opened Charles Cros's envelope, which placed Cros as the first inventor of the phonograph, ahead of Thomas Alva Edison[3] by a few days. Indeed, Edison filed his patent on December 24 for a recording device with tin foil lying on a cylinder.

**Figure 1.3.** *Edison's phonograph*

On March 11, Edison presented his phonograph to the *Académie des Sciences* and on April 22 to the press. On April 24, 1878, the Edison Speaking Phonograph Company was founded with the purpose of promoting, improving and selling this new invention. Edison envisioned new possibilities such as the "sound letter", and even the premise of the speaking clock, etc. He was never short of ideas and, ever the good businessman, he wanted the phonograph to be present everywhere, in homes and companies.

In 1886, Alexander Graham Bell[4] and Charles Tainter[5] filed a patent for a vertical recording device, the *graphophone*, whose medium was a cardboard cylinder covered with wax and natural kerosene (ozocerite or fossil wax). Together, they created the American Graphophone Company and the Volta Graphophone Company.

---

3 Thomas Alva Edison, 1847–1931, American inventor and scientist, founder of General Electric. Prolific, he filed more than 1,000 patents.

4 Alexander Graham Bell, 1847–1947, Scottish Canadian scientist, inventor and engineer, naturalized American. Winner of the Hughes Medal in 1913.

5 Charles Sumner Tainter, 1854–1940, American engineer and inventor.

The term vertical engraving deserves an explanation, because this technique, as well as its competitor, lateral engraving, remained at the center of discussions about the different types of phonographs for several decades.

Vertical recording was the first to be used for cylinders. The tip of the stylus engraves the disk deeply, while the depth of the groove is modulated by the sound. The grooves are perfectly equidistant.

The disks with lateral engraving have a furrow of constant depth, whose position varies on the right and on the left of the unmodulated furrow. The spacing between the grooves is, therefore, not constant. These disks are read with a very sharp needle that wears out quickly; it is thus necessary to change it regularly under penalty of irremediably destroying the disk.

It was quickly realized that lateral engraving gave much better results in terms of sound quality; however, the Pathé company used vertical engraving until the end of the 1920s. The reading was done with a stylus provided with a spherical sapphire; the advantage, compared to the needles used in the lateral system, was that there is very little wear.

**Figure 1.4.** *A Pathé sapphire pick-up for a vertically engraved disk. For a color version of this figure, see www.iste.co.uk/reveillac/recording.zip*

The reading styli are not compatible with each other; a needle cannot read a vertically engraved disk, as it would destroy the groove, and a sapphire would slide on a laterally engraved disk, scratching its surface.

The year 1887 was important; the Edison phonograph was going to be perfected, the tin foil cylinder was abandoned in favor of a cardboard medium on which fossil wax was deposited (invented by Bell the year before). Many inventors like

A.G. Bell, G. Bettini[6] and E. Berliner[7] saw the possibility of abandoning the cylinders, which remained too fragile and of a very limited conservation in time, to replace them by a disk.

On September 26, 1887, E. Berliner filed a patent concerning the recording on disk and invented a process of duplicating the disks by galvanoplasty[8]. Edison created the Edison Phonograph Company.

**Figure 1.5.** *Emile Berliner and his first phonograph (source: Wikipedia)*

In 1888, Edison presented a phonograph equipped with an electric motor. He also turned to the solution of the wax-coated cylinder (Bell and Tainter's graphophone), abandoning the tin foil which had a much lower performance in terms of sound quality.

That same year, Berliner founded the Gramophone Company.

6 Gianni Bettini, 1860–1938, Italian inventor who worked in the fields of sound and cinema.

7 Emile Berliner, 1861–1929, German-American engineer and inventor.

8 A technique for applying a metallic deposit dissolved in a liquid to the surface of a metal object.

The Americans Louis T. Glas and William S. Arnold presented the first public phonograph (juke-box[9]), the "*multiphone*", on November 22, 1889, in San Francisco. It was a multi-cylinder phonograph with a coin-operated machine.

Between 1890 and 1892, the idea of archiving sound content was born, whether it was music or narrative content. It was also the beginning of the first duplication devices that could engrave new rolls from a mother cylinder. The process used *pantographs*; the most efficient ones could copy up to 50 rolls simultaneously. The engraving speed was much slower (about eight times slower) than the reading speed, in order to obtain a duplicate of the best possible quality. The master cylinder needed to be re-recorded, on average, for every hundred copies.

In 1893, the French watchmaker Henri Lioret developed a new type of cylinder. It was made of celluloid and was considered unbreakable.

**Figure 1.6.** *The Columbia Graphophone Grand. For a color version of this figure, see www.iste.co.uk/reveillac/recording.zip*

9 The term juke-box emerged in the 1930s in the USA.

The year 1896 marked the birth of the Pathé Frères, the company of brothers Charles and Emile Pathé. They imported and marketed American phonographs in France. Two years later, the French researcher François Dussaud perfected the electric reading and recording of cylinders.

The format of the cylinders evolved, with their diameters increasing from 5.5 cm to 12.5 cm, which required the manufacture of a suitable machine. Columbia, which was at the origin of this new format, designed the "Graphophone Grand", capable of reading these new cylinders.

In order to keep up with the competition, Edison also had to create a machine compatible with large cylinders. Following in Columbia's footsteps, he marketed the "Edison Concert" phonograph model, which worked with cylinders 11 cm in diameter.

**Figure 1.7.** *The "Edison Concert" phonograph. For a color version of this figure, see www.iste.co.uk/reveillac/recording.zip*

It should be noted that during the following years, many different cylinder formats were created; not only did their dimensions vary (length: 106, 108, 110, 125 mm; diameter: 55, 88, 110 mm), but also the speed of rotation was necessary to their reading (100, 125, 144 and 160 rpm). The average duration of listening was two minutes and four minutes for the Edison Amberol Four cylinder (1908). There were also re-recordable rolls like those of the *Distaphone* (1903). Among the general public, there was a lot of confusion about which format and material to choose.

## 1.3. Magnetic recording

On December 1, 1898, Valdemar Poulsen[10], taking up the work of Oberlin Smith who had published, a few years earlier, an article entitled "Some possible forms of the phonograph" in the journal Electric World, filed a patent for a device called the *Telegraphone*, which used magnetic recording. In fact, this device was the ancestor of the telephone answering machine, broadcasting a message of a few seconds to the caller and making it possible for them to record and leave a message themselves. This invention was presented at the Universal Exhibition in Paris in 1900.

The medium of the sound message was a steel wire wound on a cylinder and the reading–writing head was a small electromagnet.

**Figure 1.8.** *Poulsen's "telegraphone"*

Although in principle the telegraphone was revolutionary for its time, few people showed interest in magnetic recording.

At the beginning of the 20th century, five large companies shared the market: Edison, Victor Recording Company, which was later bought by RCA[11], Columbia, Pathé and Gramophone. They manufactured and marketed the machines, engraved the cylinders, duplicated them and took care of their distribution.

10 Valdemar Poulsen, 1869–1942, Danish engineer and inventor.

11 In 1929, the RCA (Radio Corporation of America) bought the Victor Recording Company, which was, at the time, the world's largest manufacturer of phonographs.

At the 1900 Paris Exposition, we could see films, of short duration, sounded with the help of cylinders, and synchronized manually. Numerous improvements appeared, such as models of phonographs with multiple reading heads and horns to enable greater sound volume ("Multiplex" phonograph, from Columbia), the telegraph, the telephone answering machine, the luxurious "Céleste" phonograph from Pathé, etc.

**Figure 1.9.** *The Columbia Multiplex, with its three pavilions*

In 1902, the sound was automatically synchronized with the image of the projector, the "*Chronophone*", a machine developed by Gaumont and Decaux.

## 1.4. The advent of 78 rpm

Between 1902 and 1906, professionals mobilized themselves and understood that cylinders were not the best choice. They were cumbersome and fragile for some; they instead turned to the disks. Columbia abandoned cylinders in 1912 and Edison in 1929.

In 1906, Pathé, who understood that having a large catalog of various works was important to achieve commercial success, invented a device to transfer the contents of its cylinders to flat disks. This device was named the "Poisson" pantograph. With this process, the disks obtained started towards the center and had a speed of rotation of approximately 95 rpm.

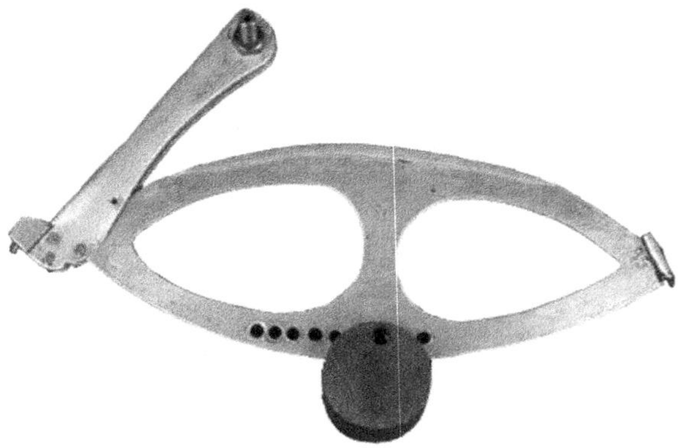

**Figure 1.10.** *Pathé's "Poisson" Pantograph. At the left end is the engraving stylus and at the right end, the reproducing sapphire. At the bottom is the weight compensation and at the top, the rotation and holding axis*

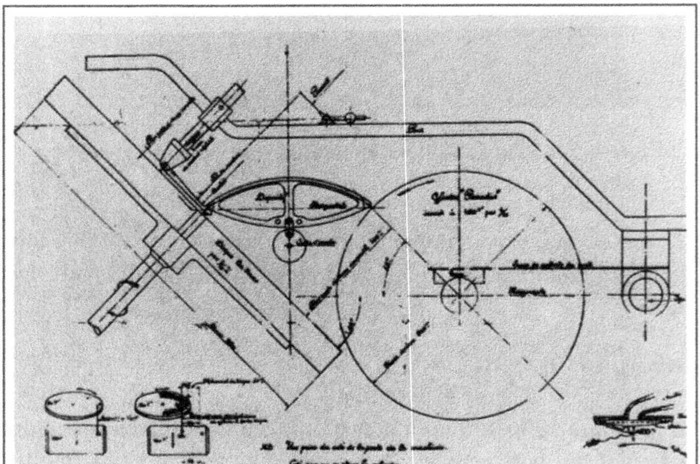

**Figure 1.11.** *Pathé's "Poisson" principle (source: La nouvelle revue du son, no. 286)*

In 1908, Columbia began to produce double-sided disks, which doubled the recording time. This was the decisive point that generated a shift from the cylinder to the disk.

**Figure 1.12.** *A Columbia double-sided 78 with its cover. For a color version of this figure, see www.iste.co.uk/reveillac/recording.zip*

From 1912 onwards, the cylinders and disks were duplicated by electroplating, a process that made it possible to produce copies of excellent quality in large quantities.

NOTE.– It is necessary to note an important point: it is the reading stylus that required needles. These were made out of steel most of the time, but other materials were also used (ivory, glass, bamboo, etc.). The classic needles wore out quickly and had to be changed often (practically after each reading) to keep the record in good condition and to have a good quality listening. The advent of tungsten steel, around 1905, enabled a longer life (about 50 readings), but it was the sapphire stylus, very widespread after 1910, which predominated by offering minimal wear, better sound quality and low degradation of the groove. The steel stylus eroded the record.

Faced with the multitude of dimensions and speeds for the cylinders, the same phenomenon appeared for flat disks; however, the speed of 78 turns was going to predominate as the reference, with a diameter between 20 and 50 cm included, generally readable by all the apparatuses.

Why 78 rpm? This was due to the technology in the USA – the synchronous motors that equipped most *gramophones*[12]; this was the name retained for the flat

12 Over time, the terms phonographs and gramophones have become confused, and today, both are used.

disk phonographs which were supplied with 110 V 60 Hz and turned at 3,600 rpm. They are connected to a speed reduction system of ratio 46, which leads to a speed of 3,600/46 rpm, that is, 78.26 rpm exactly.

At the beginning of this century, many artists were being recorded and questions of ethics were beginning to arise. Would these recordings harm their career, their image and their prestige? It is true that the quality of the recordings gave the voice a very nasal, parasitic and not very faithful sound.

In 1925, electrical recording developed rapidly with the appearance of the microphone and the quality of the recordings became much better.

The Columbia Company bought the Gramophone Company from Pathé brothers in 1928. Three years later, it became the Electrical and Musical Industries Ltd. (EMI).

From 1927, we started to find devices with an electronic amplification[13] that finally gave a significant sound volume and a good reproduction quality. In general, the apparatus is associated with a radio that shares the amplifier and the loudspeaker, with the whole apparatus being assembled in a commode-type piece of furniture. Pavilion gramophones started to disappear. The diaphragm reading head was replaced by the electric pick-up.

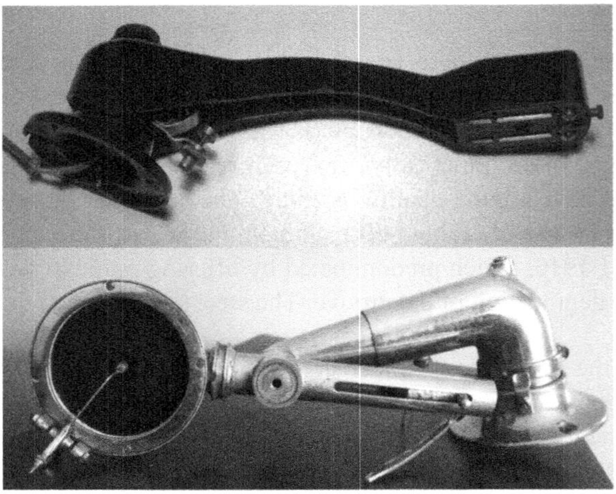

**Figure 1.13.** *The electric pick-up (top) and the membrane read head (bottom)*

13 The invention of the electronic tube, in 1906, and of the triode, by the American engineer Lee De Forest made the technologies progress very quickly, particularly those related to sound and image, with the possibility of amplifying the current modulated by a microphone or a reading head.

At the end of the 1920s, the broadcasting and military industries were finally interested in magnetic recording.

In 1931, RCA released the recordable cellulose acetate disk, which was widely used by radio stations before the advent of magnetic recording.

**Figure 1.14.** *Examples of electric pick-up phono cabinets (wireless telegraphy). For a color version of this figure, see www.iste.co.uk/reveillac/recording.zip*

The first phonograph cases that made the gramophone transportable also made their debut. They were mechanical and became electric in the mid-1930s.

**Figure 1.15.** *A mechanical phonograph case. For a color version of this figure, see www.iste.co.uk/reveillac/recording.zip*

In 1934, the German physicist Curt Stille[14], who studied magnetic recording in detail and was associated with the English company Marconi, presented one of the first magnetic recorder-players. In 1933 he created the *Blattnerphone*. The medium used is a steel ribbon that is 3 mm wide, running at a speed of 1.5 m/s, which can record a sound message of 30 minutes. The device uses three heads, one for reading, one for recording and one for erasing. It is enormous and weighs several hundred kilos. It must be said that the electronics of the time were very rudimentary and many technical problems had to be solved (constant speed, amplification of the signal from the microphone, bandwidth, etc.).

**Figure 1.16.** *Curt Stille's magnetic reader-recorder (source: www.EMTEC.co.uk)*

At the same time, a German engineer, Fritz Pfleumer[15], taking up the work of Oberlin Smith, coated a paper tape with iron particles to create the first recording tape and built a device capable of reading and recording it. Unfortunately, the sound quality was poor. Pfleumer registered the patent in 1930, but was unable to promote and market his machine.

14 Curt Stille, 1873–1957, German physicist and chemist, one of the fathers of magnetic recording and the first steel wire dictation machines (Dailygraph and Textophon).

15 Fritz Pfleumer, 1881–1945, German-Austrian engineer.

**Figure 1.17.** *Fritz Pfleumer with his magnetic tape machine*

## 1.5. The magnetic tape and the LP

It was not until 1932 that the president of the German company AEG (Allgemeine Elektricitäts-Gesellschaft) bought the rights to Pfleumer's patent and devoted part of his research to magnetic recording. Its position as a manufacturer of electrical and electronic equipment was not sufficient to ensure total control of the processes necessary for magnetic recording. Therefore, AEG called on the IG Farben company, which specialized in chemistry, to finalize what would become the magnetic tape.

It was 6.5 mm wide and made of acetate. It was coated with *carbonyl iron* powder, a pure iron of more than 97%. The scrolling speed was 77 cm/s.

In 1934, AEG presented to the public, its first *tape recorder*[16], the K1, a self-contained, fully integrated device with its own amplifier and loudspeaker.

As early as 1939, AEG worked on a magnetic head with two air gaps that made it possible for two tracks to be recorded on the same tape, but it was not until 1942 that the first stereophonic tape recorders appeared.

16 The name *Magnetophon* was originally a trademark of AEG and IG Farben for a magnetic tape recorder.

**Figure 1.18.** *The AEG K1 tape recorder (the letter K stands for "Koffer", the suitcase). For a color version of this figure, see www.iste.co.uk/reveillac/recording.zip*

From 1946 onwards, magnetic recording on tape became widespread and began to attract the general public in 1950.

The first 78 rpm records were made of *shellac*, a naturally occurring plastic material discovered by Fred Gaisberg[17]. It is a substance obtained from the secretion of an Asian insect to which slate powder and wax are added. The first disks made from this material appeared in 1898 and production stopped in 1948. During World War II, the importation of shellac was interrupted and records were produced in polyvinyl chloride, which was less resistant to the action of gramophone needles.

In 1946, Peter Goldmark[18] took up the idea for Columbia, in a different form, and the company patented an entirely new, lighter *polyvinyl chloride* record. One of the advantages of this material, which is a thermoformable plastic, is that reproduction costs are reduced.

For this new support, the size of the groove is reduced and the speed of rotation has been lowered to 33 1/3 revolutions per minute, which offers a longer reading time for each side, up to 23 minutes. The diameter[19] is set at 12 inches or 30 cm. The

17 Fred Gaisberg, 1873–1951, musician and one of the first sound engineers and producers.

18 Peter Carl Goldmark, 1906–1977, German-Hungarian engineer, naturalized American.

19 The diameter of 30 cm is the most common; however, microgrooves of 25 cm also exist.

reduction of the speed is possible because the vinyl lets the stylus slide with much less friction than the shellac, thus having the effect of drastically reducing the background noise and increasing the range of transmitted frequencies.

Columbia marketed its product under the name of *microgroove* and *long play* (LP).

It should be noted that the support force that was 100 g for the 78 rpm, played on a non-amplified gramophone with needles, passed between 10 and 30 g for a reading on a pick-up with an electronic amplifier to reach less than 2 g with the microgroove disk with a reading point of 15 µm.

The *single* or 45 rpm record (17.5 cm in diameter with a large hole in its center) arrived in 1946 and sounded the death knell for the 78 rpm records, which were still around for a few years.

NOTE.– The larger hole was intended to facilitate its use in jukeboxes that developed considerably since the 1920s. There are, however, 45 rpm with a small hole, identical to the 33 rpm.

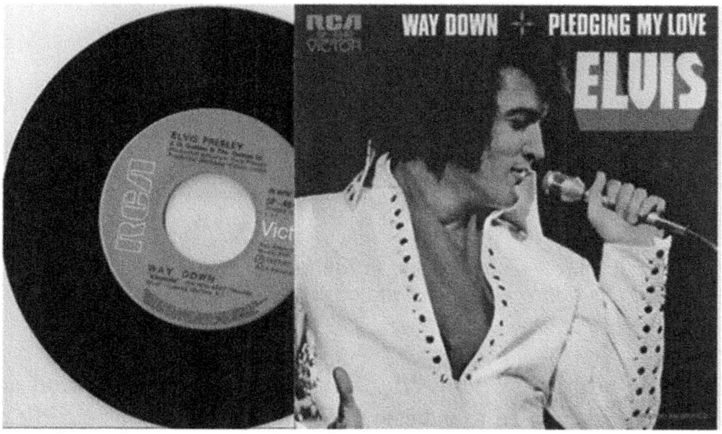


**Figure 1.19.** *A 4 (rpm) disk and its cover. For a color version of this figure, see www.iste.co.uk/reveillac/recording.zip*

At the end of the 1940s, Deutsche Grammophon introduced the variable pitch, which tightens the grooves during low-intensity passages in the recording process. This process makes it possible for the duration per side of a record to be substantially extended.

At the beginning of the 1950s, several companies marketed record players for the general public, also known as *electrophones*. They were presented in the form of a case whose lid almost always contained the loudspeaker. The base of the apparatus supported the turntable, of reduced dimension, approximately 20 cm in diameter, and its articulated reading arm carrying a cell that was generally piezoelectric. Most of them had several playback speeds (33, 45, 78, and sometimes 16 rpm), which were mechanically selectable. In the base, there was also a pre-amplification device (with tubes and then with transistors a few years later), which offered the user a volume control and sometimes a bass-treble correction.

**Figure 1.20.** *An electrophone from the 1950s. For a color version of this figure, see www.iste.co.uk/reveillac/recording.zip*

The LP gave birth to the term *Hi-fi* (High Fidelity), which spread among the general public. It is supposed to indicate that equipment with this qualification is of a superior quality, bringing a sound reproduction very close to the original. Contrary to what one might think, there is no Hi-fi standard.

In 1958, the LP improved with the appearance of stereophonic records developed in 1931 on 78s by the engineer Alan Blumlein[20], who had called the process *binaural sound*. The stereophonic recording condenses the two sound channels into a single groove. It is a compromise between vertical and lateral recording.

20 Alan Blumlein, 1903–1942, English engineer who specialized in electronics related to telecommunications and sound recording. He filed more than 128 patents during his career.

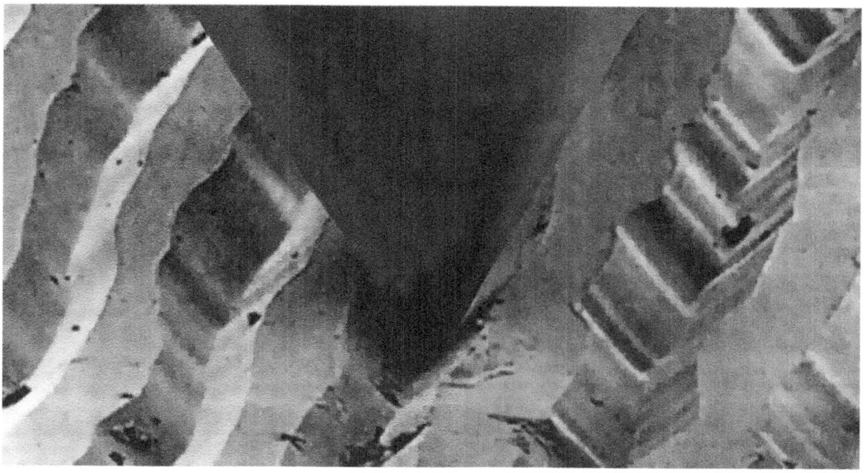

**Figure 1.21.** *Pick-up and groove of a stereophonic LP*

In 1951, the first miniature 1/4 inch magnetic tape recorder, the Nagra I, appeared. It was designed by the Polish engineer Stefan Kudelski[21], who did not stop improving it. He built the subsequent models II, III, IV and IV-S (stereo) between 1953 and 1971.

**Figure 1.22.** *The Nagra I news reel-to-reel tape recorder with miniature lamps. For a color version of this figure, see www.iste.co.uk/reveillac/recording.zip*

21 Stefan Kudelski, 1929–2013, engineer and designer of the high-quality Nagra recorders, dedicated to cinema and radio. Founder of the Kudelski company.

## 1.6. 8-track cartridges, mini-cassette and Trimicron

The *cart machine* introduced by Collins Radio arrived in 1959 in the USA. It was used by the radio to broadcast jingles or commercials.

**Figure 1.23.** *An 8-track tape and its internal structure. For a color version of this figure, see www.iste.co.uk/reveillac/recording.zip*

A few years later, in 1962, it turned into a 4-track version, then in 1964, an 8-track version. It became the *8-track cartridge* or 8-track tape or *stereo 8*, which was very popular from the end of the 1960s to the beginning of the 1980s, especially in car radios, although players, and even player-recorders, exist as equipment of a high-fidelity system.

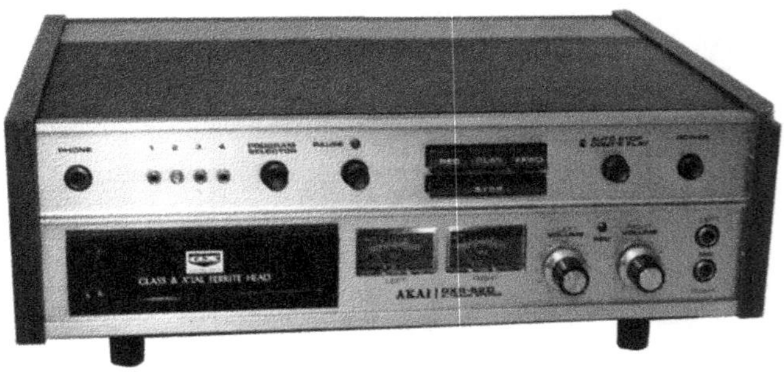


**Figure 1.24.** *The Akai GXR-82D 8-track tape player/recorder*

This cartridge is an endless tape that was invented by Bernard Cousino in 1952 that embeds a 1/4 inch magnetic tape that runs at 9.5 cm/s. The 8-track version was designed by Richard Krauss who also modified and simplified its mechanics. Its duration was 80 minutes in stereophony.

The 8-track tape disappeared in 1990, although very limited editions were still produced until 2011.

**Figure 1.25.** *A limited edition 8-track released in 2009 – Cheap Trick – "The Latest". For a color version of this figure, see www.iste.co.uk/reveillac/recording.zip*

At the beginning of the 1960s, tape recorders had made enormous progress and were very widespread. The reels had diameters of 8, 13, 15, 18 and 27.5 cm. The average recording time of a 13 cm reel is one hour, but can vary according to the speed and thickness of the tape.

**Figure 1.26.** *The Telefunken 76 (Tube) tape recorder, 4 tracks (stereo 2 X 2 tracks), speed: 4.75–9.5 cm/s, reel to 15 cm diameter, dating back to 1960. For a color version of this figure, see www.iste.co.uk/reveillac/recording.zip*

They can be mono, with 2 times 1 track, stereo with 1 times 2 tracks, stereo with 2 times 2 tracks. Tube electronics were progressively replaced by transistors, which made the manufacturing of more compact devices possible.

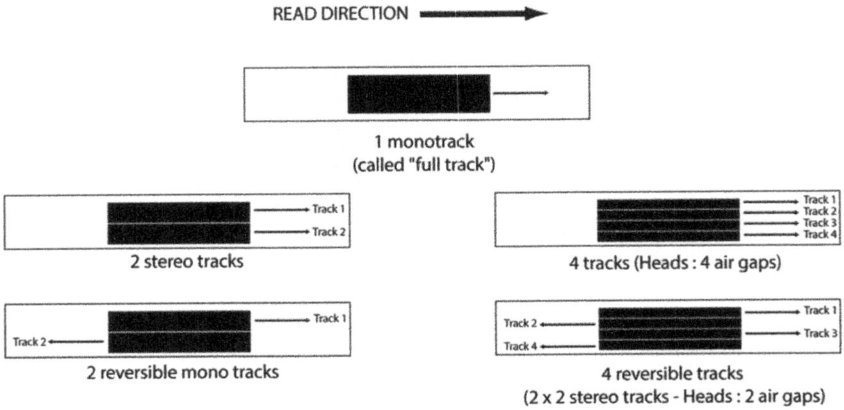


**Figure 1.27.** *The different formats of magnetic tape recording*

In 1963, the audio cassette or *mini-cassette*, abbreviated *K7*, was launched by Philips. It was used with a specially adapted tape recorder, the *mini-cassette recorder* or cassette recorder, which could be integrated into a more complex device such as a radio cassette or a high-fidelity system.

**Figure 1.28.** *One of the first mini-cassette player-recorders with its microphone, the Philips Model EL3302*

The K7 had 4 tracks, 2 stereophonic tracks for each side, since it can be turned over. Its reduced dimensions (100 x 83 x 12 mm), as well as its use in car radios and Walkmans made it an essential piece of equipment everywhere in the world until the appearance of the CD and even long after.

**Figure 1.29.** *A type I – C90 (2 X 45 min) mini-cassette from TDK. For a color version of this figure, see www.iste.co.uk/reveillac/recording.zip*

NOTE.– The mini-cassette has found variants in other dimensions such as the microcassette, of smaller dimensions, reserved for dictaphones (57 x 38 x 11 mm).

It was an element present in most of the consumer hi-fi systems and in the studios for 25 years. Its decline only came at the dawn of the 2000s, with the rise of digital players.

**Figure 1.30.** *The Tascam 122 MKII professional cassette deck. For a color version of this figure, see www.iste.co.uk/reveillac/recording.zip*

In parallel with cassette decks, manufacturers tried to combine a cassette player-recorder and a multitrack mixer in the same unit. This gave birth to small portable studios, the first of which was the Portastudio 144 from Tascam. Over the years, and still today, the term *portastudio* is used to describe a stand-alone multitrack recorder.

**Figure 1.31.** *The Portastudio 144 stand-alone multitrack from Tascam (1979). For a color version of this figure, see www.iste.co.uk/reveillac/recording.zip*

Four types of K7 existed:

– type I: normal (30 Hz to 15 kHz) – 1963;

– type II: chrome (30 Hz to 16 kHz) – 1970;

– type III: ferrichrome (30 Hz to 16 kHz) – 1975;

– type IV: metal (30 Hz to 18 kHz) – 1980.

Their duration at a scroll speed of 4.75 cm/s is 30 (C60), 45 (C90), 60 (C120) and 90 (C180)[22] minutes per side. The K7 used by the music publishers, known as "musicassettes", were of variable length according to the lengths related to the recorded works.

22 These K7 produced by TDK were known to be very fragile, with the thickness of the tape being very low.

**Figure 1.32.** *A musicassette – Metallica – "Master of Puppets". For a color version of this figure, see www.iste.co.uk/reveillac/recording.zip*

In the USA, mainly, the 4- or 8-track tape developed in parallel.

The year 1964 brought a novelty that spread in the small world of the LP, also called *stereo-mono* or *synchro-stereo*. It made it possible to indifferently read the record in mono or in stereophony. Technically, it is a stereophonic recording for which the vertical modulation amplitudes are limited to a certain threshold. The vertical component thus decreased ensured the good functioning of the monophonic reading head for a reading of the horizontal engraving (only existing in the case of a monophonic recording).

In 1971, the *tetraphony* or *quadraphony*, *quadrophony*, *quadrasonic* emerged on the market; the objective was to record the atmosphere of the place of recording to restore it to the listener. This innovative technique was, however, a commercial failure because of many technical problems, including the fact that the user had to have a material that could restore the four channels, which was very expensive.

In the middle of the 1970s, the *Trimicron* developed by MDR (Magnetic Disk Recording) and invented by a French researcher, E. Rabe, tripled the possible duration of an LP, thus ensuring one hour of music per side. These 33 turns had a certain success; unfortunately, the advent of the CD (Compact Disk) made it disappear quickly.

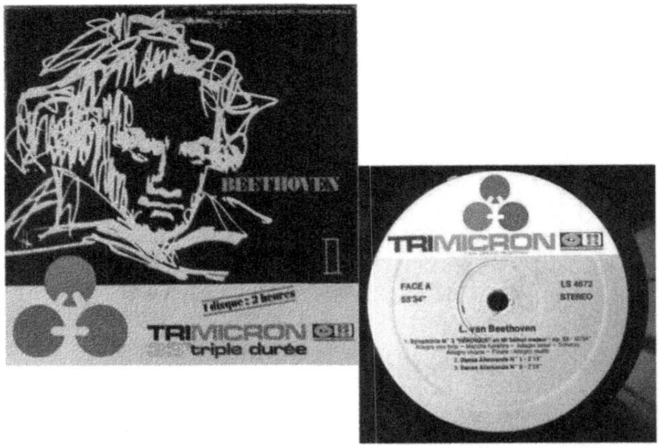


**Figure 1.33.** *A trimicron disk from MDR. For a color version of this figure, see www.iste.co.uk/reveillac/recording.zip*

**Figure 1.34.** *A 24-track 2-inch multitrack (Studer A800 – MKII), a standard of excellence in the studios of the 1970s and 1980s. For a color version of this figure, see www.iste.co.uk/reveillac/recording.zip*

Tape recorders also saw their performance increase. While individuals still confined themselves to 4-track stereophonic tape recorders, working at speeds between 2.38, 4.75, 9.5 and 19 cm/s, in recording studios, it was not uncommon to find multi-track equipment working on magnetic tapes, whose widths were ¼, ½, 1 and 2 inches. They were capable of recording 2, 4, 8, 16, 24, or even 32 tracks in parallel, whose tape ran at 19, 38, 76 cm/s, or even 114 cm/s, with the sound quality increasing with the speed.

## 1.7. The compact disk and the advent of digital technology

At the beginning of the 1980s, the digital era made its entrance, relegating analog processes to the background, even though it took several years for this new technology to become established.

In 1979, Philips and Sony Corporation decided to work together to produce a new digital medium for music to replace LPs. The specifications imposed significant improvements, including:

- an increase in listening time;
- a reduction in background noise;
- an improvement in sound quality;
- a medium less fragile than the LP (scratches, dust, etc.);
- a smaller, lighter and less cumbersome medium;
- an extended frequency range of 20 Hz to 22 kHz;
- frictionless reading, no stylus tips;
- indexing of the ranges to move quickly from one to the other;
- an accelerated reading option for tracking;
- a time display for each track.

All of these technical characteristics were listed in the *Red Book.*

Philips designed the medium and the playback system, while Sony defined the digital format and the related software devices. The final medium was the CD (Compact Disk), with a diameter of 120 mm and a thickness of 1.2 mm, covered with a thin layer of aluminum, offering a maximum duration of 74 minutes with a 16-bit coding and a sampling frequency of 44.1 kHz.

**Figure 1.35.** *The first CD released by Sony, Billy Joel – "52nd Street". It came with the purchase of a Sony CD player. For a color version of this figure, see www.iste.co.uk/reveillac/recording.zip*

On August 11, 1982, Philips and Sony announced the release of their new product and the production started quickly. The first commercialized players were the CD100 from Philips (August 17, 1982) and the CDP-101 from Sony (October 1, 1982).

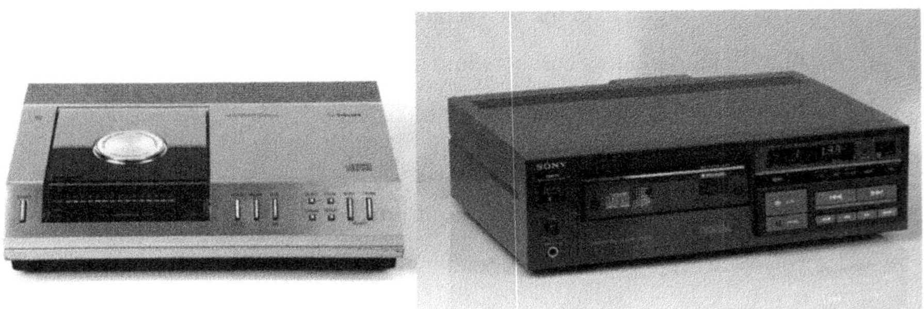


**Figure 1.36.** *The first two commercial CD players, the Philips CD100 (left) and the Sony CDP-101 (right)*

The first portable player was released in November 1984; it was the Sony D50 (Discman). The era of the Walkman[23] was starting.

23 Walkman is a trademark of Sony that has become an antonomasia over time.

**Figure 1.37.** *The first portable CD Player, Sony D50 "Discman"*

## 1.8. Digital technology is essential

While the CD was the new medium of the 1980s, in the recording studios, digital technology was also making its entrance. New machines emerged to make digital recordings. The DASH (Digital Audio Stationary Head) standard, presented by Sony, brought a new digital audio tape format in 1982. Other manufacturers like Studer, Teac and Matsushita also took this option.

NOTE.– To be accurate, it should be noted that at the same time, another standard, named *ProDigi*, also exists. It was designed by Mitsubishi. Its popularity remained much less than the DASH system. However, it was quite popular in Nashville, and was employed in the studios dedicated to country music. ProDigi fell into disuse in the mid-1990s with the advent of the first DAWs (Digital Audio Workstations) and direct-to-disk on hard disk.

**Figure 1.38.** *Studer D827 Mk II (DASH 48 tracks) with its remote control console. For a color version of this figure, see www.iste.co.uk/reveillac/recording.zip*

With DASH, you can record two tracks on a ¼ inch tape and up to 24 or 48 tracks on a ½ inch tape. The PCM (Pulse Code Modulation) encoded data is stored linearly and parallel on the tape; this makes it possible for you to work as with analog magnetic tape, by cutting it with a razor blade to make edits. The sampling frequencies supported by DASH are 44.1 and 48 kHz, with a resolution of 16 bits.

**Figure 1.39.** *Sony PCM 3324 S digital multitrack (DASH 24 Tracks). For a color version of this figure, see www.iste.co.uk/reveillac/recording.zip*

At the end of the 1980s, the transition to digital technology shook up everything related to sound at a frantic pace, both for the general public who abandoned their LPs and vinyl turntables in favor of CD players, and for professionals who sent their analog machines to the scrap heap.

Innovations are multiplying and new standards are appearing all the time; the audio industry is looking for itself. While for the average listener, the compact disk became the reference, for recording studios and radios, there was no stability. In this boiling environment, the DAT (Digital Audio Tape) cassette made its entry in 1987, with the objective of replacing the audio cassette (mini-cassette) by offering a duration of three hours in a digital format 16 bits, 48 kHz (superior to the CD). However, it did not win the patronage of individuals, as the readers-recorders were very expensive. On the other hand, the DAT system made it possible to produce perfect digital clones by duplication, which forced manufacturers to introduce limitations on certain models of consumer DAT decks.

It is, therefore, limited to professionals who are satisfied to find a compact solution (73 x 54 10.5 mm) with high audio quality to save their recordings in a digital format without compression.

NOTE.– Be careful not to confuse the DAT audio cassette with the DDS (*Digital Data Storage*) cassette, which has the same format but is intended for digital data storage. Even if a DSS cassette is accepted without any problem in a DAT player-recorder, you should know that DSS tapes are not compatible and there is a risk that the playback heads may permanently deteriorate.

**Figure 1.40.** *A studio DAT player-recorder – Sony PCM 2800*

The DAT player-recorder has a rotating playback head and a helical scan of the tape, similar to a VCR, and the magnetic tape is 3.81 mm (0.15 inch) wide. The tapes can have durations of 15–180 minutes[24] and are not reversible.

The history of the DAT tape ended in December 2005, but many professionals still use it today.

As previously mentioned, the DAT equipment is very expensive; to compete with Sony, Philips and Matsushita invented a new medium in 1992, the digital compact cassette or DCC, for the general public with less expensive devices. It was to supplant the mini-cassette in this period oriented towards digital. It is not reversible; only an auto-reverse device reads the opposite side, but it keeps a compatibility with the mini-cassette that made it possible for the DCC player to read a mini-cassette as a DCC cassette as well.

24 The 180 minute tapes, because of their thickness, were very fragile. A 120-minute tape is 60 m long.

**Figure 1.41.** *A DAT cassette (73 x 54 10.5 mm), on the left, compared to a mini-cassette (100 x 83 x 12 mm), on the right. For a color version of this figure, see www.iste.co.uk/reveillac/recording.zip*

**Figure 1.42.** *A DCC cassette*

The DCC recorder uses a fixed magnetic head and linear recording. The cassettes have a maximum duration of 105 minutes. One of the fundamental differences between DAT and DCC is that for the latter, the audio data is compressed[25] to ensure a greater recording capacity. For this reason, the DCC cassette was often decried as the poor man's DAT cassette by purists.

25 Compression is ensured by a coding called PASC (Precision Adaptive Subband Coding).

**Figure 1.43.** *The Matsushita RS-DC8 technics DCC player-recorder*

It is necessary to recognize that like the DAT, the DCC did not meet the success expected by the users, although Walkman were even manufactured.

This beginning of the 1990s still reserves many other digital surprises, after the DAT cassette and the DCC cassette, a new product, the ADAT digital audio tape was announced in 1991 at the NAMM convention[26] in Anaheim.

The American company Alesis designed this new system. The ADAT devices, which were also called *Blackface,* had 8 tracks and could be synchronized between them, offering the possibility of managing 128 tracks simultaneously (with 16 machines in parallel). The first models worked at 44.1 kHz on 16 bits, the later models accessing 48 kHz on 20 bits. Backward compatibility was assured.

**Figure 1.44.** *The first-generation ADAT Blackface player from Alesis. For a color version of this figure, see www.iste.co.uk/reveillac/recording.zip*

26 National Association of Music Merchants, the music industry's largest trade show.

The medium used was a Super VHS (S-VHS) cassette identical to that used in consumer VCRs.

**Figure 1.45.** *An S-VHS tape for use with an ADAT recorder. For a color version of this figure, see www.iste.co.uk/reveillac/recording.zip*

To compete with Alesis, Tascam, one of the divisions of the Teac Corporation also released its own machine, the DA-88, a digital multitrack recorder, in 1993. This machine uses Hi-8 video cassettes as a recording medium and is capable of recording eight tracks or more by combining several devices. The resolution is 16 bits; later, it will be increased to 24 bits (DA-78HR and DA-98HR) for a sampling frequency of 44.1 or 48 kHz. The recording format is called DTRS (*Digital Tape Recording System*).

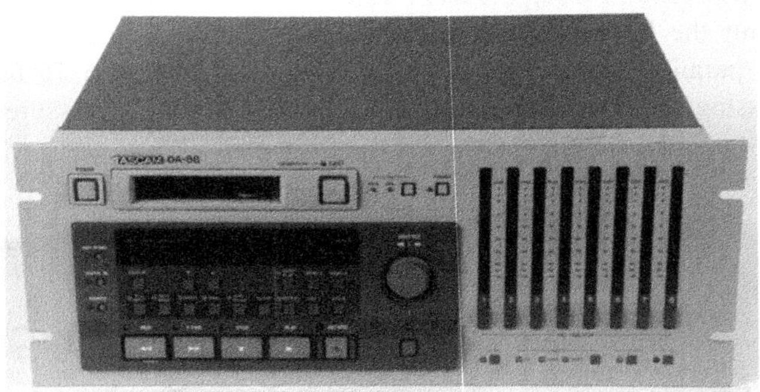

**Figure 1.46.** *The Tascam DA-88 digital player-recorder. For a color version of this figure, see www.iste.co.uk/reveillac/recording.zip*

A 120-minute Hi-8 cassette provides 108 minutes of audio recording.

The DA-88 was replaced by several models with different characteristics, DA-38 and DA-98, and a model edited by Sony, the PCM 800.

**Figure 1.47.** *A 90 minute Hi-8 cassette (95 x 62 x 15 mm). For a color version of this figure, see www.iste.co.uk/reveillac/recording.zip*

In 2012, the DTRS format was definitively abandoned and even while some afficionados still use it, magnetic tapes are increasingly difficult to find.

Alesis, Tascam and Sony had their digital recorders, but Akai was not left behind and conceived the A-DAM format with its huge player-recorder DR1200 on an 8 mm video cassette, which came out on the market in 1988. Its resolution was 16 bits and its sampling frequency was 44.1 or 48 kHz. It offered 16 minutes of recording on a 60 minute cassette at 44.1 kHz or 14.5 minutes at 48 kHz.

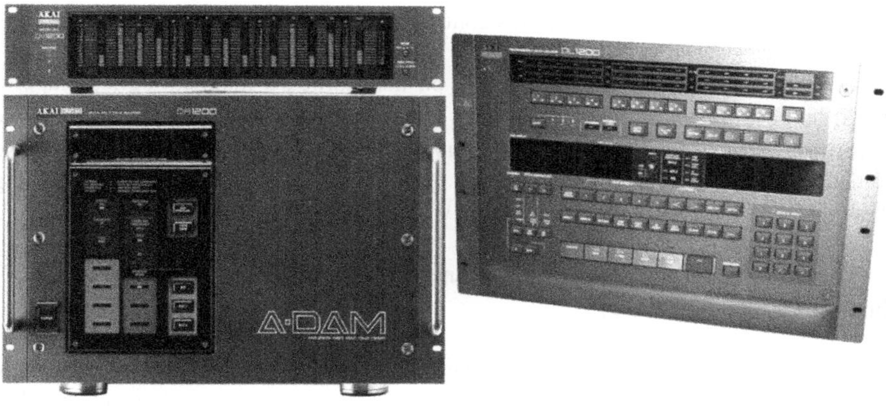


**Figure 1.48.** *The imposing Akai DR1200 reader-recorder with its bank of vu-meters (DM 1200) and its locator (DL 1200). For a color version of this figure, see www.iste.co.uk/reveillac/recording.zip*

## 1.9. Hard disk recorder and minidisc

In the mid-1990s, other types of devices appeared, *hard disk recorders*, multi-trackers that record directly on the hard disk. Many manufacturers have their own models, Alesis, Teac, Sony, Fostex, Studer, etc. They all have different specificities with a number of tracks that can generally go from 4 to 24, with a resolution of 24 bits, at 44.1 or 48 kHz.

**Figure 1.49.** *The Fostex D80 hard disk recorder – 8 tracks (1996). For a color version of this figure, see www.iste.co.uk/reveillac/recording.zip*

These machines and their features continued to evolve until the mid-2000s.

In 1992, Sony launched the *minidisc* on the market, a magneto-optical support that it had developed a few years before. The coding used is the ATRAC (*Adaptive Transform Acoustic Coding*), a data compression system that stores 80 minutes of audio recording. This duration increased to five hours with the advent of the Long Play player in 2002, and then to 45 hours with the high-density minidisc in 2004.

The minidisc was presented as a small disk integrated in a plastic case of dimensions 68 x 72 x 5 mm.

In the West, the minidisc never created the expected craze, the material being expensive, both for the minidiscs themselves and the players. The advent of the MP3 format, at the beginning of the 2000s, slowed down the expansion of the minidisc among the general public, even though its quality was clearly superior.

**Figure 1.50.** *Tascam MX-2424 – 24-bit, 96 kHz, 24-track hard disk recorder (2004) with its RC2424 remote control. For a color version of this figure, see www.iste.co.uk/reveillac/recording.zip*

**Figure 1.51.** *A minidisc. For a color version of this figure, see www.iste.co.uk/reveillac/recording.zip*

**Figure 1.52.** *The Tascam MD-801R MKII minidisc player-recorder. For a color version of this figure, see www.iste.co.uk/reveillac/recording.zip*

The year 2013 marked the end of production of audio systems dedicated to the minidisc.

## 1.10. Microcomputer, direct-to-disk and DAW

Around the year 2000, dedicated multitrack recorders became rarer, with the democratization of microcomputers and the first *direct-to-disk* editing and recording software.

The first recorders available on microcomputers appeared at the beginning of the 1990s, following one of the first tools, *Sound Designer*, intended for editing a music file, most often from a sampler (Akai S900, Korg DSS-1, Ensoniq Mirage, E-Mu Emulator, etc.). It worked on Apple Macintosh (Macintosh Plus, the then Macintosh II).

With the rapid evolution of micro-computing, Digidesign created *Sound Tools* in 1989 together with *Sound Designer II*, an evolution of the first version. Sound Tools was a software that worked with a dedicated electronic card (in Nubus[27] format), still on Apple Macintosh. It was able to acquire from a DAT and managed AES/EBU[28] and S/PDIF connections. Its resolution was 18 bits and it could manage two channels.

The year 1991 saw the advent of the first *Pro Tools* system, based on a modified version of *Deck*, the first digital multitrack recording software, based on a

27 Nubus is a parallel computer bus developed at MIT (Massachusetts Institute of Technology).

28 AES/EBU and S/PDIF are standards that define digital audio transmission interfaces available on many professional audio devices since the late 1980s.

microcomputer, developed by OSC, a small Californian company, Pro Tools based on the *Audiomedia* and *Sound Accelerator* cards designed by Digidesign.

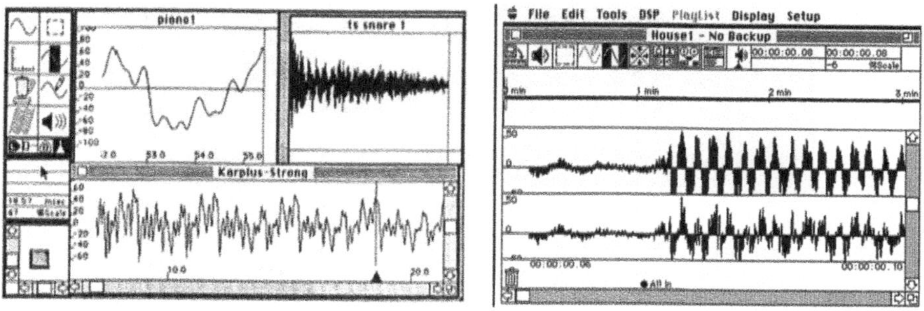


**Figure 1.53.** *Digidesign sound designer (left) and sound tools (right)*

**Figure 1.54.** *The Audiomedia II board in Nubus format for Apple Macintosh. For a color version of this figure, see www.iste.co.uk/reveillac/recording.zip*

In 1993, Pro Tools II, a completely rewritten version of the first version and correcting its main flaws and weaknesses, arrived on the market. It offered 16 tracks to which plugins could be added and which worked in real time. Pro Tools II became one of the first digital audio workstations (DAW) used by all professionals.

Version 2.5, released in 1994, used a new multiplexing technology. The same year, Digidesign merged with Avid.

In 1995, Pro Tools III offered 48 tracks and two new interfaces named 882 and 888; moreover, a series of plugins were provided by default: equalizers, compressors, delays, reverbs and modulations.

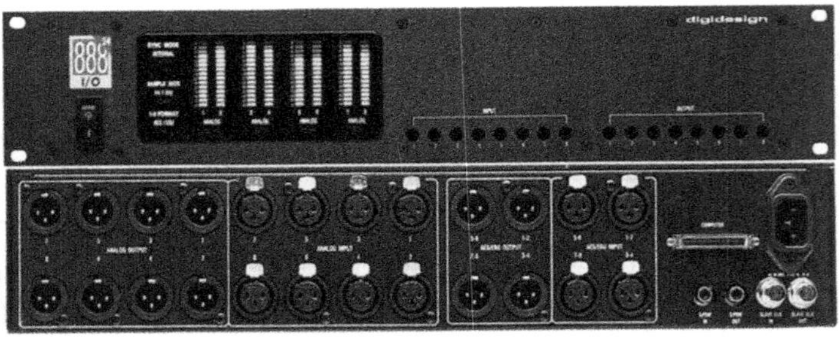


**Figure 1.55.** *The 888 interface for Pro Tools III. For a color version of this figure, see www.iste.co.uk/reveillac/recording.zip*

In 1996, Apple abandoned the Nubus bus within these machines to switch to the PCI bus, which made the PC and Macintosh world converge. Pro Tools was able to work in the Microsoft Windows environment in 1997.

In the following years, Pro Tools underwent significant improvements and consolidated its position in professional studios, while opening up to the general public (Pro Tools LE – 1999).

**Figure 1.56.** *Pro Tools 2020 version. For a color version of this figure, see www.iste.co.uk/reveillac/recording.zip*

Today, Pro Tools is still present, but it is no longer alone and no longer has superiority; many other DAWs are present on the market at more competitive prices and with much better features. The general public is no longer interested in Pro Tools, and professional studios are increasingly moving away from it. The competition is tough, and even if its position remains strong, Pro Tools is running out of steam.

My previous presentation on DAWs focused on Pro Tools because it is the precursor in this field, but many other software packages appeared in the early 1990s. It would be difficult to give a history of each one and I prefer to list them in Table 1.1, which shows the chronological order of their appearance. Many of these programs have evolved over time and are still around today. The list presented is far from being exhaustive; I have concentrated above all on the tools with multitrack audio-editing capabilities.

| Year | Name | Publisher | Comments |
|---|---|---|---|
| 1990 | Studio Vision | Opcode | Macintosh, PC |
| 1991 | Cubase audio | Steinberg | Macintosh |
| 1994 | Music Maker | Magix | PC |
| | Emagic | Logic Audio | Macintosh |
| 1995 | Magix | Samplitude | PC |
| 1996 | Twelve Tone Systems | Cakewalk Pro Audio 4 | PC |
| | Cubase VST | Steinberg | Macintosh, PC |
| 1997 | Fruity Loops (FL) | Image-Line | PC |
| | Canam Computers | Quartz Studio | PC |
| 1998 | Acid | Sony | PC |
| 1999 | Pro Tools 5 | Digidesign | Macintosh, PC<br>Integrates MIDI |
| 2000 | Nuendo | Steinberg | Macintosh, PC |
| 2001 | Live | Ableton | Macintosh, PC |
| | Cakewalk Sonar | Roland | PC |
| 2003 | Traktion | Raw Material Software | Macintosh, PC |
| | Audition | Adobe | Macintosh |
| 2004 | Live 4 | Ableton | Macintosh, PC<br>Integrates lunch |
| | Logic Pro 6 | Apple | Macintosh |
| 2005 | Garage Band | Apple | Macintosh |
| | Reaper | Cockos | PC |
| 2006 | Samplitude Pro 9 | Magix | PC<br>Integrates MIDI |
| 2007 | Logic Studio | Apple | Macintosh |
| 2009 | Studio One | Presonus | Macintosh, PC |
| 2010 | Bitwig Studio | Bitwig | Macintosh, PC-Linux |
| 2013 | Logic Pro X | Apple | Macintosh |
| 2017 | Digital Audio performer | MOTU | Macintosh, PC |

**Table 1.1.** *Some digital audio workstations (this list is far from exhaustive)*

## 1.11. To conclude

This presentation has shown that recording devices have not stopped evolving since the middle of the 19th century, with a certain boom during the transition to digital technologies at the beginning of the 1990s. I have deliberately not mentioned the last 20 years, which have not created a real technological revolution. The principles have remained the same; certainly the performances have improved, the storage capacities have grown, the processing speed and the computing power have increased 10-fold and the size of the machines has reduced (today a digital recorder of excellent quality fits in a pocket), but, finally, the principles have remained the same.

It is important to note a contradiction, or rather a countercurrent, which has been grafted onto the evolution of technologies, the return in force of LPs (vinyl) since 2005, as shown in Figure 1.57.

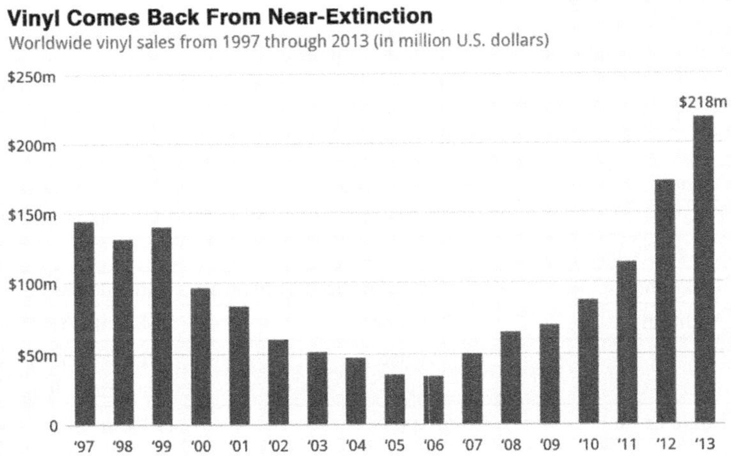


**Figure 1.57.** *Evolution of the vinyl record market worldwide (source: IFPI.com). For a color version of this figure, see www.iste.co.uk/reveillac/recording.zip*

Could it be the nostalgia, the sound coloration, so different from digital, the fashion, the object itself? It's hard to say, personally I think it is a combination of all of these and the numbers are growing; you can even find vinyl in supermarkets.

On the recording side, the industry has to face this revival and press the good old tape again. For independent studios, there are even autonomous burning devices that burn blank vinyl by the unit because a demand has been established.

**Figure 1.58.** *The T560 vinyl engraving bench of the German manufacturer Souri's Automaten. For a color version of this figure, see www.iste.co.uk/reveillac/recording.zip*

Whatever the case, recording remains a never-ending adventure that will never cease to amaze us.

# 2

# The Voice

The voice, the vector of speech, can also be considered as a wind instrument. Talking, shouting, whispering, humming and singing, this way of communicating using the voice, is one of the characteristics of human evolution.

The voice is the result of complex interactions within our body. Thousands, even millions of variants, exist, giving each individual a different tone.

In this chapter, I will describe the vocal apparatus that we use to emit sounds and how it works. I will then approach the spoken or sung voice, the breath, the tessituras, the vocal quality and the characteristics attached to the timbre.

## 2.1. The vocal apparatus and its functioning

A set of anatomical structures located in the thorax, neck and head must coordinate in order for us to make sounds. Everything depends on our breathing; without it, no sound could be formed.

The process begins in the airways with an inspiration. The diaphragm stimulates the rib cage to expand and the air reaches our lungs and bronchial tubes, which form the lower airways.

The air pressure will then vibrate the vocal cords, located in the *larynx*, at which point the breath reaches the *phonation* stage. The more the air pressure generated increases, the more the vibratory frequency of the vocal cords increases and the note emitted rises towards the treble. Control of this frequency controls the pitch of the sound emitted.

---

For a color version of all the figures in this chapter, see: www.iste.co.uk/reveillac/recording.zip.

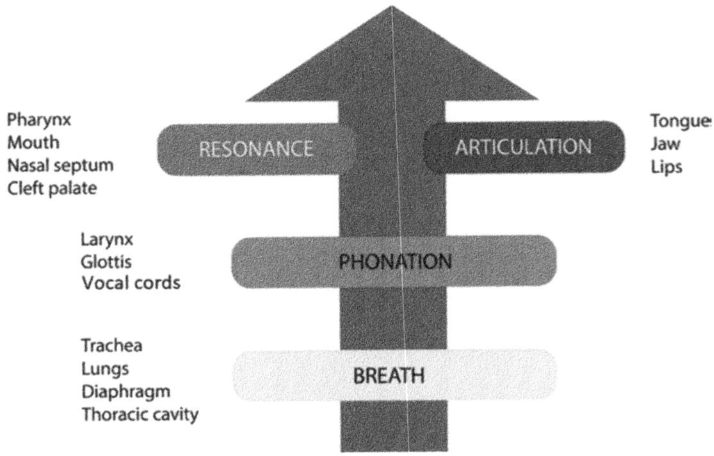


**Figure 2.1.** *The stages of sound creation in a human being*

The *trachea*, formed by the *tracheal rings*, the lungs and the diaphragm are the air reservoir that carries the voice.

Located in the neck, the larynx is mobile and suspended by an elastic muscular system. Associated with the glottis and the vocal cords, it forms the sound. It is here that vibrato is generated.

After the *phonation* stage will come, in parallel, the stages of resonance and articulation. The oral cavity (the throat and the oral space), the nasal septum and the soft palate will define the tone of the voice and ensure the major part of the vocal identity. They will also determine the amplification.

The tongue, the jaw and the lips will ensure the articulation by modifying the resonances in order to control the formation of consonants and vowels.

I would like to come back to the important role of the larynx, which is the central organ of phonation. This cartilaginous element ensures the connection between the *pharynx* and the trachea. It is covered by a mucous membrane, and its task is to stop foreign contents from going towards the trachea or the lungs thanks to a reflex which implements the *epiglottis*. It is also a sound regulator that works in perfect harmony with the vocal cords.

The larynx unconsciously moves when swallowing, speaking or singing. In humans, it can be clearly recognized by the presence of the *Adam's apple*.

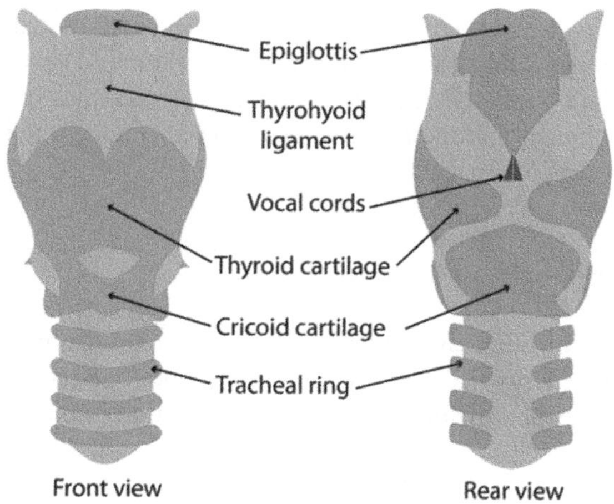


**Figure 2.2.** *The different elements of the phonation chain*

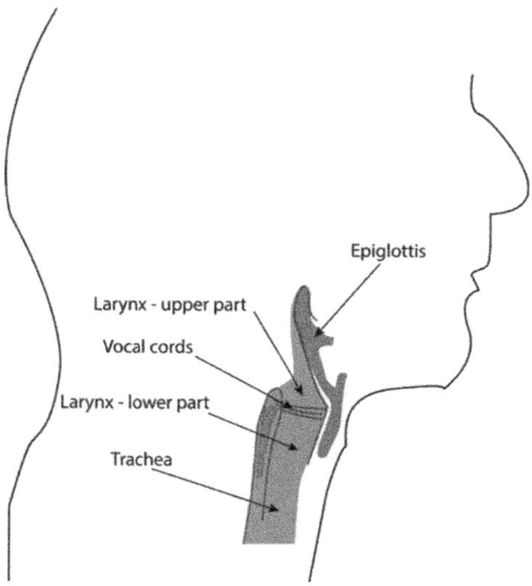


**Figure 2.3.** *Larynx and vocal cords, the key elements of the voice's timbre*

The larynx can be raised or lowered voluntarily, which significantly alters the voice. Most singers learn to keep the larynx in a comfortable position, at the right height, in order to be as relaxed as possible during their performance. Breathing combined with a controlled use of the diaphragm to provide a consistent volume of air that limits the upward movement of the larynx is also a fundamental element.

The voice is fragile; however, a performer must be able to maintain an accurate and optimal voice for what can be long periods of time. The oral cavity, which serves as a sounding board, determines the main characteristics of the vocal timbre. It is in this environment that the *harmonics* or *formants* appear and develop, which differentiate the voice and make it unique.

## 2.2. Voice and breath

Building a voice relies, as I mentioned earlier, on breathing and the breath, which provides the energy necessary to build the sound. Phonation, which allows us to formulate syllables or a melody, is a secondary function of the breath.

The act of emitting a phonatory breath is initially built around an inspiration during which the air enters the lungs, then the bronchi, secondary bronchi, bronchioles and pulmonary alveoli. During the phonatory phase, the air travels in the opposite direction and enters the larynx at a flow rate, speed and pressure adjusted according to the voice to be produced.

In fact, a murmur can be broken down into three modes, upper thoracic, abdominal and mixed.

The upper thoracic mode, or high breathing, is the result of the lowering of the rib cage due to the action of the intercostal muscle. The chest rises during inspiration and expiration results in an inward tilting of the sternum and a collapse of the rib cage. This breathing mode is used for conversational speech and does not involve any control of the pressure generated by exhalation.

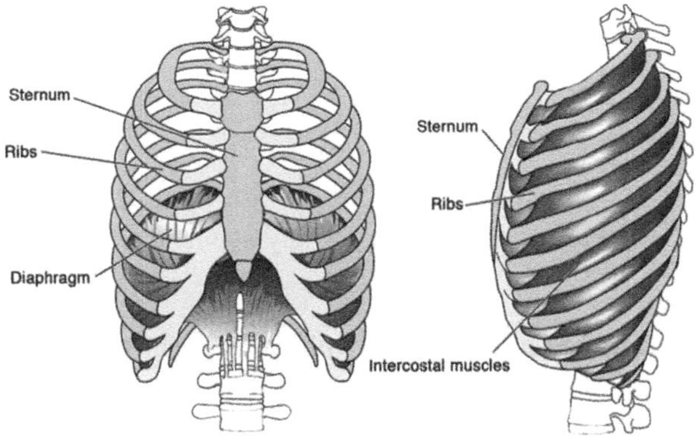


**Figure 2.4.** *Thoracic cage, sternum and intercostal muscle*

The abdominal mode works the oblique and transverse muscles, which has the effect of retracting the abdominal wall, causing the diaphragm to be pushed upwards and the rib cage to be lowered, thus tightening the rib cage laterally.

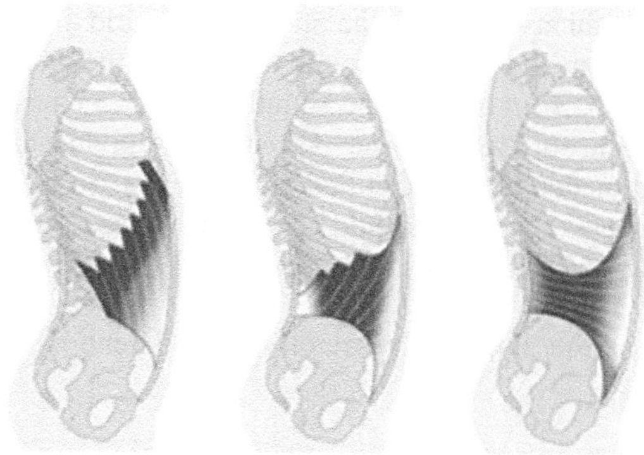


**Figure 2.5.** *The oblique and transverse abdominal muscles*

The abdominal breath moves the upper part of the sternum forward and upward. Through it, the voice is projected as when we sing or declaim a text. The contraction of the abdominal muscles makes it possible to control the respiratory pressure.

The mixed mode combines the thoracic and abdominal breaths, which together give the voice a greater intensity. The use of this mode remains unusual because it requires an unnatural operation.

Abdominal breathing, also called diaphragmatic breathing, is the basic starting point for singing.

## 2.3. Song and speech

Singing means articulating and modulating your voice; it involves automatisms that affect the cortex as well as the ears and the larynx.

Singing accentuates vowels, while conversational speech relies on consonants.

Previously, we have seen the physiological and anatomical aspects related to the voice, but we must not neglect the intellectual and psychological aspects. Singing, declaiming a text or even doing a voice-over require reflection and the implementation of feelings that must be transmitted to create emotion in the listener.

Vocal expression is complex and difficult to control; for example, the posture is a physical criterion that influences the quality of the speech or the singing.

Singing a melody, reading a text aloud, is telling a story; a good performer invests and projects themselves into an imaginary world in agreement with the narration; these emotions must shine through to enhance the musical writing and the meaning of the text. It is up to you, the sound engineer, or possibly a third party, such as a vocal coach, to reassure your performer by giving them confidence, by providing your comments, always distilled and succinct, so as not to drown them under a host of elements that they should correct. It is necessary to progress slowly and make successive corrections in small steps.

The keywords of the singer or narrator remain intention and motivation; the listener must feel it.

Technology can do a many things, make corrections, but will never add the emotional aspects that are the foundation of a good vocal performance.

Being a perfectionist is not always the best choice; sometimes, you have to accept a few flaws or slight mistakes rather than getting a perfect recording, but one that is devoid of feelings and sensations.

## 2.4. Frequency, intensity and timbre

These three characteristics are fundamental indicators of the voice.

The spoken voice has a usual frequency that determines an average pitch that is always the same for a given person. This average pitch is different for children, women and men. The differences between individuals are linked to anatomical and physiological factors and also social, ethnic, family, etc. factors. When we leave the spoken voice to turn to singing, commentary or reading, the average frequency is different and is generally higher, towards the treble.

Variations of frequencies are permanent during speech; they constitute the *intonation* or the melody.

We must take into account the linguistic system. In some languages, especially in Asia, the tone is very important in terms of semantics and, therefore, understanding. In the West, the tone is also a determining factor, a simple statement can be transformed into a question simply by the intonation. These intonation patterns are attached to people who speak the same language.

It is also necessary to integrate a more psychological factor. The same voice can make one feel their joy, their sadness, their astonishment, their anger, their annoyance, etc. by its simple melodic variation.

Voice intensity can vary on a very wide scale, from a whisper (20 dB(A)) to a shout (110 dB(A)). As with frequency, a spoken voice has an average intensity used in normal conversation.

During a conversation, the intensity fluctuates greatly and depends on the intonation; some syllables or words are emphasized, others are attenuated.

The timbre, which is the result of what happens in the naso–buccal–pharyngeal cavity, is determined by the physiology of the individual. Two types of timbre are determined: the *vocal's timbre* and the *extra vocal's timbre.*

The first varies according to each vowel. It basically characterizes acoustic specificities common to all. It makes it possible to identify vowels regardless of who is speaking.

The second, the extra-vocal timbre, also called the *vocal timbre*, varies according to the individual. It depends on initial elements formulated by the larynx. Above 2.8 Khz, the sound spectrum is not very discriminating in the understanding of vowels but it gives all its color to the voice.

## 2.5. Voice and range

While every human being speaks and sings using the same phonation chain, there are fundamental differences between the voice of a woman, a man and a child and each of them can also have specificities.

In classical singing, voices are classified by *tessitura*. This is the register of sounds going from the lowest to the highest, these sounds being homogeneous with the same volume and the same timbre.

In this range, there is a comfort zone where the performer can express themselves naturally and with ease.

Eight types of voices have been defined, each defining a range:

– *soprano*;

– *mezzo-soprano*;

– *alto*;

– *tenor*;

– *baritone*;

– *bass*;

– *countertenor*;

– *contralto*.

Soprano (C3–C5) is the highest range for a woman's or child's voice. Its register is comparable to that of the violin. There are three categories of soprano: light, lyric and dramatic. Voices in this range are light and airy, but can also be very powerful.

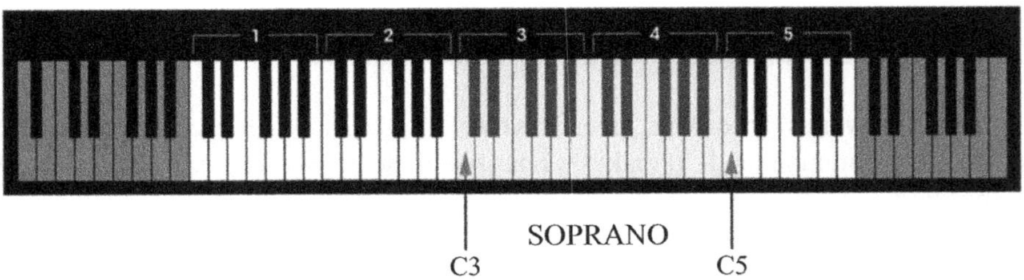


**Figure 2.6.** *The soprano range*

Mezzo-soprano is a mostly female voice that can be compared to the register of the oboe. As with the sopranos, there are three types, light (A2–C5), lyric and dramatic (FA2–A4). The power is generally identical to that of the sopranos with more limited high notes. These are the most common female voices and are suitable for many types of music such as jazz or variety.

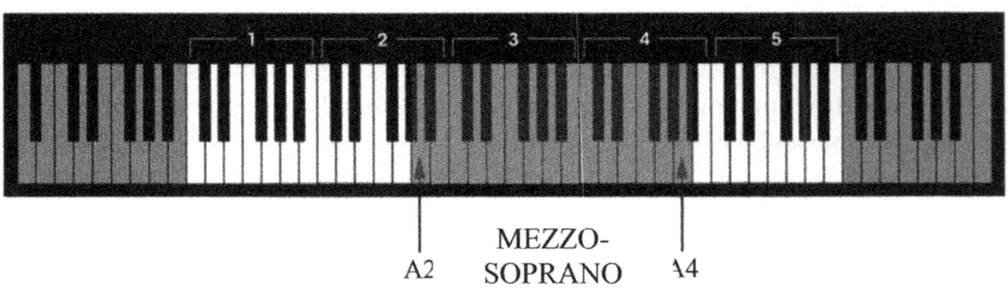


**Figure 2.7.** *The mezzo-soprano range*

Alto (F2–F4), also called contralto, is the lowest female voice that covers the range of a clarinet. It is a warm voice that gives a lot of emotion. It is often associated with mature women.

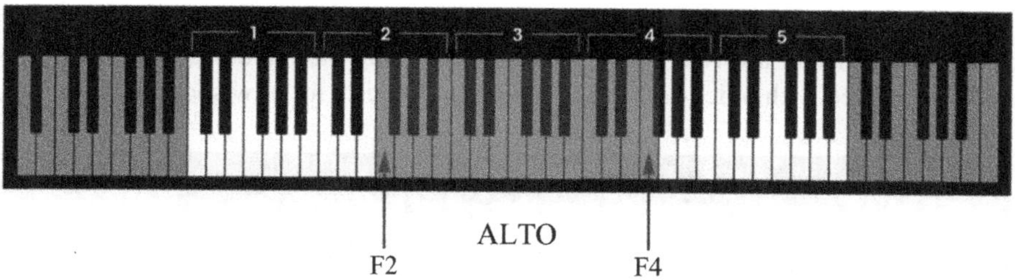


**Figure 2.8.** *The alto range*

Tenor is a range that is similar to the register of the trumpet. It is the highest male voice and the most known by the general public. Three categories exist: light (C2–D4), lyric and dramatic (C2–C4). They are often powerful, warm and intense, evolving with dexterity throughout their range.

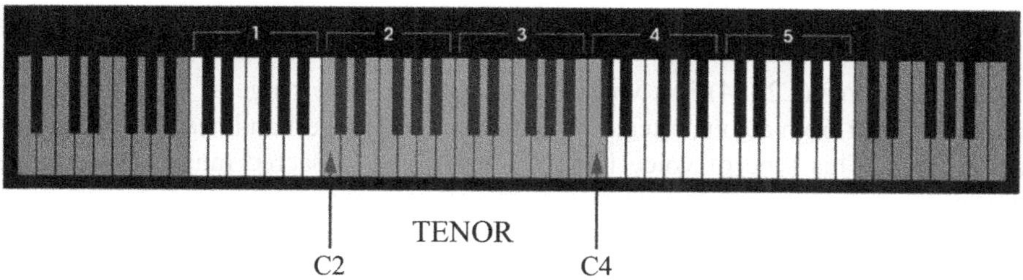


**Figure 2.9.** *The tenor range*

Baritone is a medium–low male voice that has a register comparable to the French horn. There are three main categories, light (B1–A3), lyric (A1–G3) and dramatic (G1–G3). There is a secondary category including the lyric–light tenor and the *lirico spinto* tenor. The baritone is often the principal male voice at the heart of the opera. It is one of the pillars of tragic works.

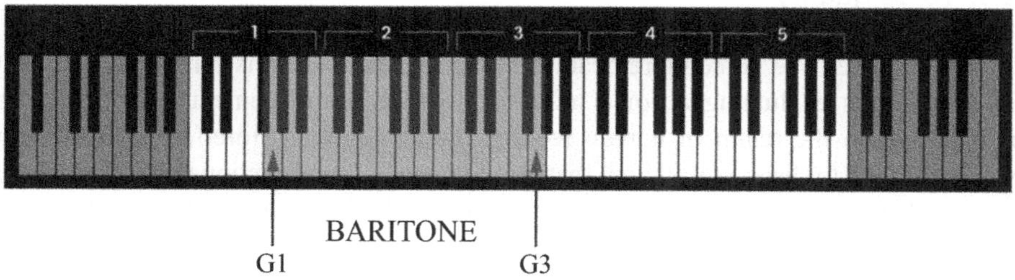


**Figure 2.10.** *The baritone range*

Bass (E1–E3) is the lowest male voice, equivalent to the register of the bassoon or trombone. There are two types of bass: the singing bass and the deep bass. It is often associated with older people and naturally carries a lot of wisdom in its timbre.

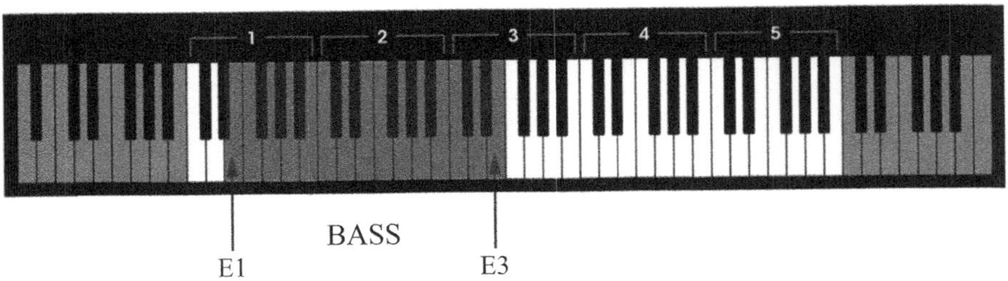


**Figure 2.11.** *The bass range*

The six preceding tessituras are very common, unlike the *haute-contre* and the countertenor, which are much rarer.

*Haute-contre* (C2–F4) is a voice that shares the same range as the tenor voice, but is different from a countertenor. It is built around a chest and head voice without any vibrato. This type of voice is very often found in sacred music and baroque music.

Countertenor is a male voice based on a head voice or falsetto. It is the male equivalent of a female soprano (also called *sopranist*), alto (also called *altist*) or counter-alto (*contraltist*) voice with much more power. Like the *haute-contre*, baroque music and sacred music are often very fond of it.

NOTE.– The indications of tessitura in the form of intervals of notes are simply there as an indication. Variations may occur depending on the references and notation used thoughout the history of music.

## 2.6. Voice quality

What makes one voice more beautiful than another? The answer to this question is difficult, as evaluating a voice must take into account elements that are often impalpable, in the realm of human feeling and perception. Moreover, to qualify a voice is also to listen to it and that implements other functions, like listening, independent of the voice itself. We can also add the influence of the environment near the performer, the culture and the objectivity, which can distort the judgment of the listener.

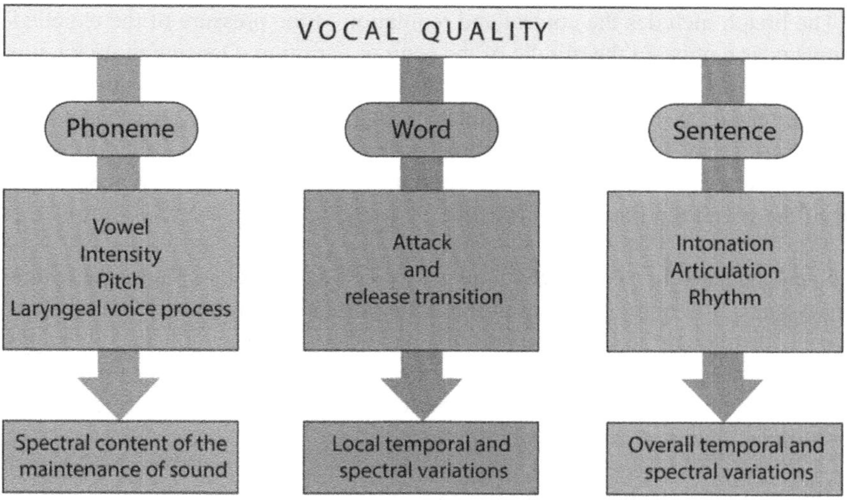


**Figure 2.12.** *Concept of voice quality*[1]

For the listener, the qualitative approach can be defined into two phases: the first in which they identify the type of performer and the second in which they take into consideration qualitative parameters related to this type.

These parameters can be located at the *phoneme*[2], word or sentence level.

Vocal quality remains a general notion which can finally be defined as what differentiates two vocal chains with the same lexical content. Parameters such as pitch, rhythm, intonation, intensity, articulation, attack, etc. can be retained, to which we can add more extensive and common notions such as linguistics, acoustics or physiology.

## 2.7. Characteristics of the vocal timbre

It can be interesting to be able to characterize a voice; in fact, we could rather say to analyze its timbre with correlated parameters which are common to all.

The breath, the *raucity*, the ornaments, the formant, the *nasality*, and the brilliance are the parameters that can be retained to which we can add the emotion felt.

1 Source: Garnier, M., Henrich, N., Dubois, D., Castellengo, M., Poitevin, J., Sothropoulos D. (2005). *Étude de la qualité vocale du chant lyrique.*

2 A phoneme is the smallest distinctive unit that can be isolated by breaking down a spoken chain.

The breath includes the control and regulation of the pressure of the expelled air. It appears as a noise in the middle of the song or narration. Depending on the type of music or the effect sought for a voice-over, it can be considered as a nuisance or, on the other hand, as a stylistic enrichment bringing authenticity. In Western lyrical music, one seeks to eliminate it, vocal purity being the main criterion. In other styles, such as jazz, rap, hip-hop and certain traditional music, on the contrary, it is part of the sound rendition that is sought.

Raucity is the characteristic of a hoarse voice, that is, containing noise within the vocal signal. It is quite common in the male population and much less common in women. It is thought to be due in part to a distortion of the sound wave at the glottis. Specialists have defined two parameters named *jitter* and *shimmer* which are at the origin of this noise. The jitter alters the periodicity of the fundamental frequency and the shimmer creates a variation of amplitude.

Ornaments are techniques present in many voices that are mostly related to the type of music being sung. They can be present or absent in the original voice but can be accentuated to create additional effects. The best known and most common is the vibrato, which modulates the sound of the fundamental note in frequency and amplitude between 4 and 8 Hz. If the frequency is higher, the effect becomes different and is similar to a tremolo. Another ornament, the *portamento*, makes it possible to link two or more notes, going through the whole range of the frequency spectrum between the first and the last notes.

The formants characterize the timbre of the voice. They are determined by the harmonics, especially between 2 and 4 kHz, which bring an important spectral richness. The formants bring a coloration that characterizes the voice and makes it recognizable and unique. Depending on the individual, they are more or less present and when a voice has very weak formants, the voice is duller and breathier.

Nasality is caused by a lowering of the soft palate or a contraction of the pharynx that brings resonances in the nasal cavities. Professionals define it by the term *twang*, an English onomatopoeia used to describe the sound of a bow string vibrating after an arrow has been shot. Nasality is a modification of the formants in terms of amplitude and on certain frequencies that can vary towards the top or bottom of the spectrum, particularly on the pronunciation of vowels.

Brilliance is one of the parameters of timbre that psycho-acousticians know well. It is linked to a difference in amplitude between the maxima of two spectral zones: 0–2 kHz and 2–4 kHz. Brilliance can be summarized as a contribution of timbre in the high frequencies as a result of the appearance of additional harmonics.

## 2.8. Conclusion

The voice is a wonderful instrument and, according to their culture, everyone can attribute a singing or speaking aspect to it. Each person has a very different spontaneous voice whose pitch, timbre, intensity and expressiveness can be evaluated. In a singing phase, we can add additional parameters, such as the tessitura, the rhythm and the tempo, associated with a certain style of singing with certain conditions of execution, environment, context, etc.

For singing, the choice of style leads to the choice of voice. Lyrical singing remains very different from singing in contemporary music. The functioning of the body overall remains the same, but the ear is educated differently, the ideals are distinct, the commercial stakes and places of performance are also different.

Depending on the style, the voice can be amplified or not. Indeed, the first condition remains being audible for the listeners. Listening live to an unamplified performer is a very rare situation.

The means of broadcasting that have become more democratic over time mean that today we listen to much more recorded music, of all styles, than to live music, and both listeners and performers have adapted.

# 3

# Microphones

Today, most of us carry a microphone without really knowing it. Our smartphones necessarily incorporate one of these devices, and it should be noted that the advent of the telephone was fundamental in the invention and evolution of the microphone. By metonymy, the piezoelectric and electromagnetic transducers are also called microphones.

We could start by asking ourselves why we chose the term microphone to designate this device. It can be broken down into two parts from the Greek (*mikros*), the affix "micro", which means small, and the word "phone" (*fônè*), which means sound or noise. Microphone can, therefore, be translated as "small sound", which is not without meaning, since this device is intended to process small sound signals.

The role of a microphone is to transform mechanical energy, sound, into electrical energy, an audio signal.

## 3.1. A little history

The term microphone was first used by the English physicist, Sir Charles Wheatstone[1].

He was one of the first physicists to understand that sound is transmitted via waves that travel through a gaseous, liquid or solid medium and to create devices capable of amplifying sound to transmit it from one place to another.

---

1 Charles Wheatstone, 1802–1975, English physicist and inventor. He was responsible for the stereoscope, the wheatstone bridge, the electric telegraph and the concertina.

**Figure 3.1.** *Sir Charles Wheatstone*

In 1861, Johann Philipp Reis[2] designed a device that converts a sound into an electrical signal that can be transmitted via an electrical wire to a second device, which converts it back into an audible sound signal.

**Figure 3.2.** *Johann Philipp Reis*

The Reis device is considered to be the first microphone in existence, in the sense that the term is used today.

It consisted of a diaphragm[3] (a membrane made of pig's bladder or parchment) stretched over a wooden frame attached to a small box to which a horn allowing the sound to be captured was connected. Above the diaphragm, two brass strips, perpendicular to each other, are attached to the box. The highly flexible of the two is in contact with the membrane and the other rigid one, is in contact with the first one through the intermediary of a drop of mercury. When speaking into the horn, the sound waves make the diaphragm vibrate, which in turn communicates the vibration to the flexible brass strip that crushes the drop of mercury to a greater or lesser extent, thus proportionally modifying the resistance induced by the contact between

2 Johann Philipp Reis, 1834–1874, German teacher, scientist and inventor.

3 In this book, I often use the terms membrane and diaphragm. There is no confusion; in my opinion, they are equivalent.

the two metallic conductors. An electric signal, running through the two lamellae, undergoes this variation and is then sent to the receiver (loudspeaker).

**Figure 3.3.** *Reis's telephone with its transmitter (microphone), left, and receiver, right (source: Deutsches museum). For a color version of this figure, see www.iste.co.uk/reveillac/recording.zip*

Unfortunately, Reis's invention has often been forgotten in history and the microphone is often attributed to Alexander Graham Bell[4], Emile Berliner[5] or even Thomas Edison[6].

In 1876, Bell and Elisha Gray[7] designed the water microphone. It consists of a metal cup filled with a solution of water and sulfuric acid, which makes the liquid conductive. A membrane acting as a diaphragm is stretched at the end of a horn. In the center of this membrane, a metallic needle is fixed. One end of the needle is dipped in the acid solution. When a sound hits the membrane, the needle vibrates. An electrical signal travels through the needle to pass through the liquid solution and is then recovered on the cup. The vibration of the needle brings about a change in

4 Alexander Graham Bell, 1847–1922, Canadian-American scientist, engineer and inventor, best known for the invention of the telephone, for which Antonio Meucci is officially recognized today as the true inventor. Winner of the Hughes Medal in 1913.

5 Emile Berliner, German engineer and inventor, naturalized American, known above all for his invention of the gramophone and the flat disc with horizontal engraving, which replaced the cylinders of the phonograph conceived by Thomas Edison.

6 Thomas Edison, 1847–1931, prolific American inventor who filed more than 1,000 patents in the fields of telegraphy, sound, cinema and electricity.

7 Elisha Gray, 1835–1901, American inventor.

resistance in the conduction that transforms the sound waves into a change in electrical current.

NOTE.– A controversy still exists today about this discovery. It concerns whether Bell or Gray invented the telephone together or independently, or whether Bell took over Gray's ideas. In any case, it was Bell who first filed the patent for the microphone.

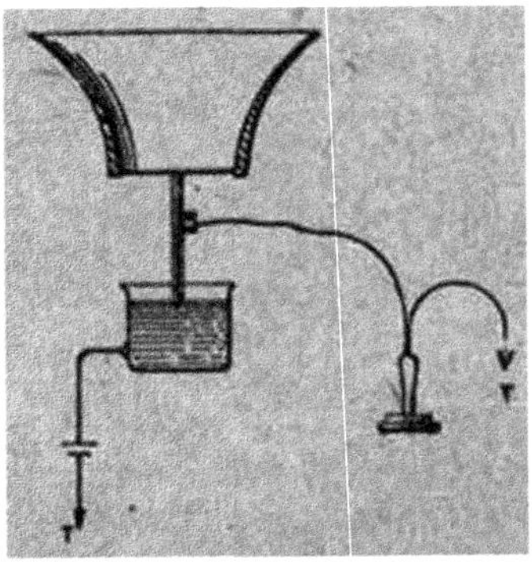

**Figure 3.4.** *The water microphone of A. G. Bell and E. Gray*

Over the months, the process was improved. The final patent filed by Bell in 1876 shows a system based on the variation of an electric current through a coil.

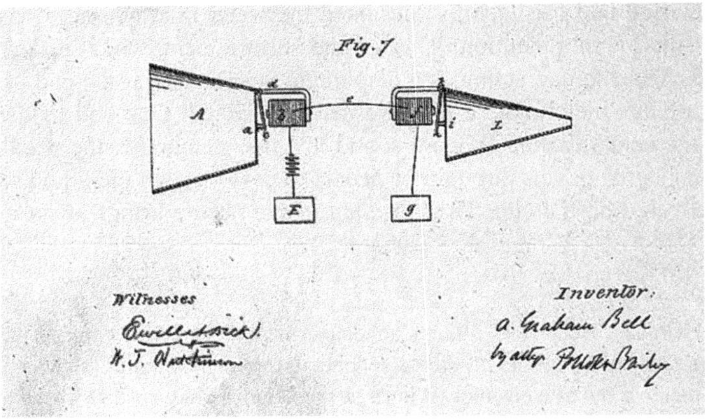


**Figure 3.5.** *One of the sketches of the telephone patent filed by A. G. Bell*

In the same year, another type of microphone was invented by Berliner and Edison. This device, called a *carbon microphone,* was much more reliable than its predecessors, and was considered to be the model that is still used as a reference today.

It is composed of two metal plates separated by carbon granules (carbon powder or carbon). One plate, at the back, is fixed and rigid, while the other, at the front, is very thin and plays the role of a diaphragm. The sound waves hitting the diaphragm exert a variable pressure on the carbon granules, which modifies the electrical resistance between the plates. By applying a continuous voltage on the plates (electrodes), we obtain a modulation of the current that translates the movements of the diaphragm.

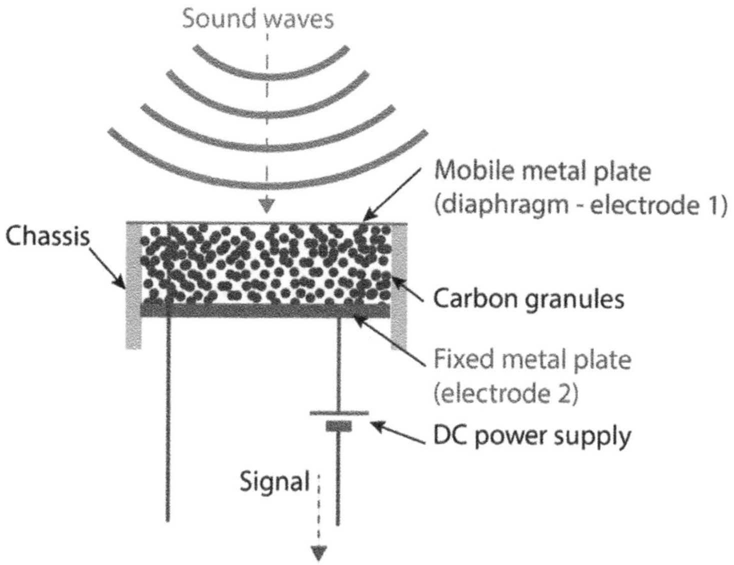


**Figure 3.6.** *Principle of the carbon microphone. For a color version of this figure, see www.iste.co.uk/reveillac/recording.zip*

On the other side of the Atlantic, in England, David Edward Hughes[8] created the same device but did not file a patent.

In 1877, Ernst von Siemens invented the dynamic moving coil microphone. A diaphragm, attached to a moving coil, moves in the field of a permanent magnet. When, under the vibration of sound waves, the diaphragm–coil assembly moves, a small current is induced.

8 David Edward Hughes, 1831–1900, Anglo-American physicist and inventor.

In 1885, transformers were invented, which brought a notable improvement to dynamic microphones, as we will see later in this chapter.

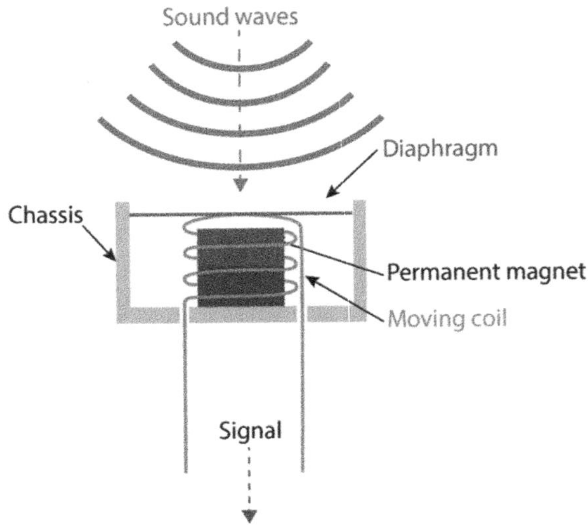


**Figure 3.7.** *Schematic diagram of a dynamic moving coil microphone. For a color version of this figure, see www.iste.co.uk/reveillac/recording.zip*

The year 1886 saw, thanks to Edison, an improvement of the coal microphone, with the torrefaction of carbon, which made it the most widely used device in most telephone systems around the world.

This microphone (*carbon-button microphone*) was used in telephone handsets until the 1980s.

**Figure 3.8.** *Charcoal microphone pad. The diaphragm can be seen on the right through the grid, and in the center of the image on the left, the electrode contact that also serves as a filler plug for inserting the carbon granules*

In 1904, J.A. Fleming[9] invented the *kenotron*, the first electronic tube (or vacuum tube or lamp), the ancestor of the semiconductor diode. It was improved by the American L. De Forest[10], in 1905, who created the *Audion*, which gave birth very shortly after, with the assistance of E.H. Armstrong[11], to the *triode*, the first electronic amplifier.

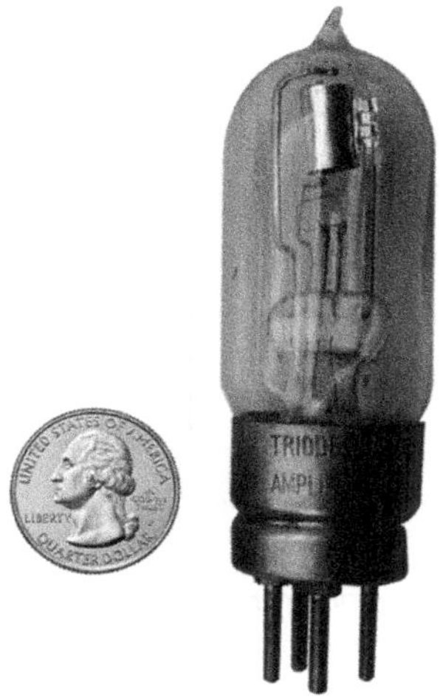

**Figure 3.9.** *A triode*

This invention played an essential role in telephony, radio, electronics, radar, photometry, electronic computers and many other fields.

In 1916, E.C. Wente[12] invented the first *condenser microphone* consisting of two parallel metal plates very close to each other, one fixed and the other mobile, of very low thickness, which acts as a diaphragm or membrane. The sound vibrations cause the mobile plate to vibrate, varying the distance from the fixed plate.

A constant charge exists by default at the terminals of the two plates. This device behaves like a capacitor, its charge varying according to the sound signals received and the distance between the two plates. The charge is then transformed into voltage and amplified to obtain a usable audio signal.

9 John Ambrose Fleming, 1849–1945, English physicist and engineer.

10 Lee De Forest, 1873–1961, American inventor who filed more than 300 patents.

11 Edwin Howard Armstrong, 1890–1954, American engineer and inventor.

12 Edward Christopher Wente, 1889–1972, American physicist and inventor who worked for Bell Laboratories.

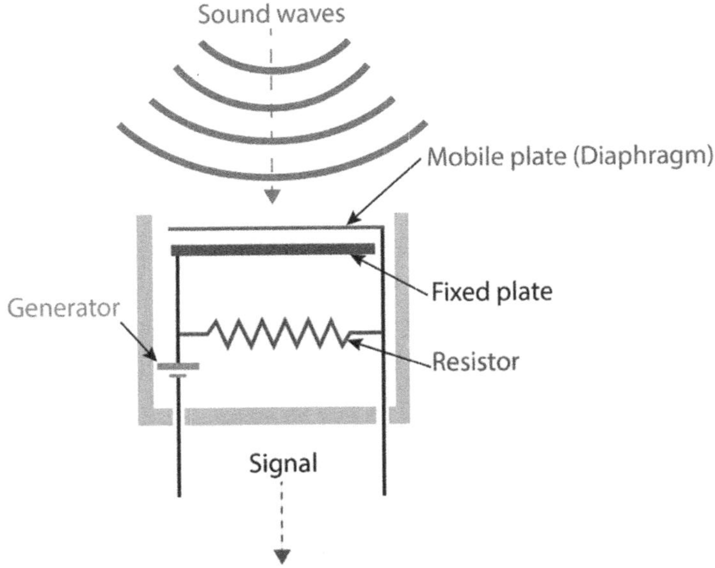


**Figure 3.10.** *Schematic diagram of a condenser microphone. For a color version of this figure, see www.iste.co.uk/reveillac/recording.zip*

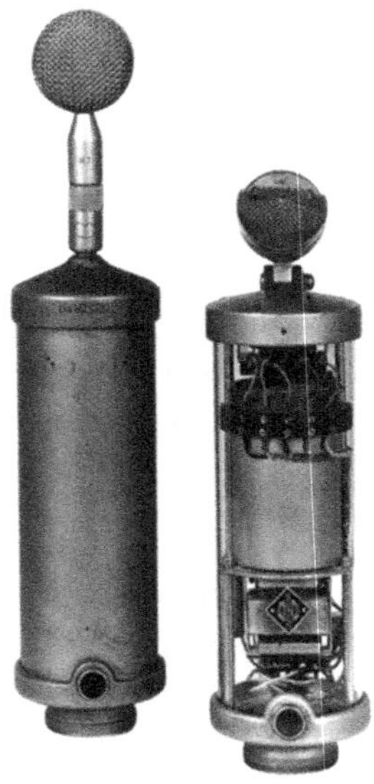

**Figure 3.11.** *The first mass-produced condenser microphone, the Neumann CVM3 (1928) – (source: en-de.neumann.com)*

The year 1917 saw the birth of the *piezoelectric microphone*, based on the work of P. Langevin[13]. It is admitted that the invention of this type of microphone is due to Alexander Nicolson, who also worked on the loudspeakers intended for phonographs. A piezoelectric crystal deforms under the action of a membrane animated by sound waves.

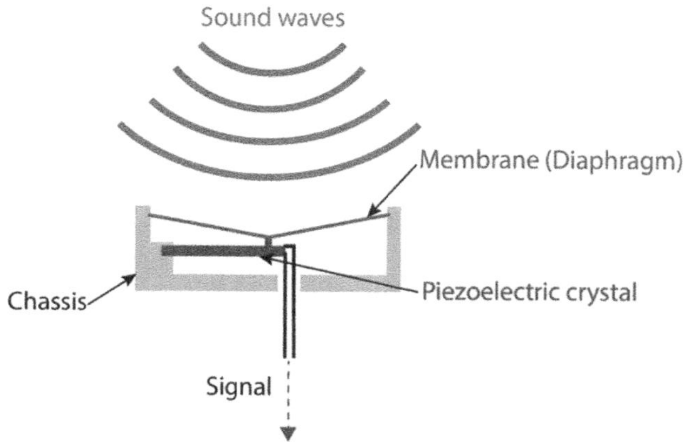


**Figure 3.12.** *Principle of the piezoelectric microphone. For a color version of this figure, see www.iste.co.uk/reveillac/recording.zip*

In 1924, W.H. Schottky[14] and Erwin Gerlach invented the *ribbon microphone*. A very thin ribbon, made of a conductive material, is suspended in a magnetic field. The sound waves that hit the ribbon, which serves as a diaphragm, move it and produce the variation of an electrical signal. It was not until the advent of high-powered magnets in the 1930s that this device was sufficiently effective. It was H. F. Olson[15] who democratized the use of this type of microphone.

The dynamic moving-coil microphone was popularized in the early 1930s.

In 1941, the first *shotgun microphones* came into existence. They consisted of a microphone to which a large tube, called an *interference tube,* was added.

---

13 Paul Langevin, 1872–1946, French physicist, philosopher and pedagogue known for his work on magnetism and Brownian motion.

14 Walter Hans Schottky, 1886–1976, German physicist and inventor specializing in electron and ion emission phenomena. Inventor of the vacuum tube with grid during his work at Siemens. He also contributed to many discoveries in semiconductor devices.

15 Harry Ferdinand Olson, 1901–1982, Swedish engineer and inventor, emigrated to the USA. He worked at RCA and filed more than 100 patents related to audio technology, especially microphones.

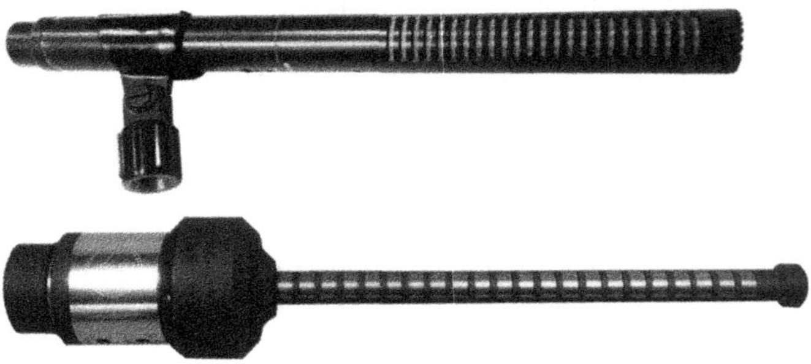

**Figure 3.13.** *Two vintage shotgun microphones: top: Sony ECM672 (1985), bottom: Electro Voice EV642 (1962). For a color version of this figure, see www.iste.co.uk/reveillac/recording.zip*

A large number of small openings are distributed along this tube. Inside the tube, a sound-absorbing material is placed, whose acoustic resistance varies along the length of the tube. One of the major problems of this system is its low sensitivity, which requires a substantial amplification of the output audio signal.

After the triode and the electronics working with tubes, in 1947, a technological revolution took place; J. Bardeen[16], W.H. Brattain and W. Shockley invented the *transistor*. This new component was used very early in microphones and in many other technological devices.

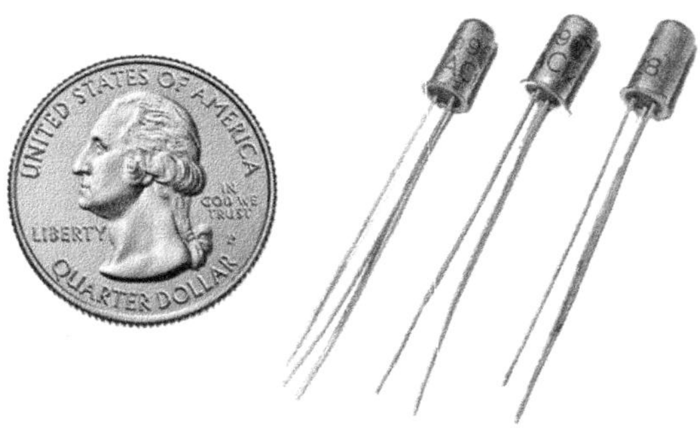


**Figure 3.14.** *Three germanium transistors. Their dimensions speak for themselves, compared to a triode*

16 John Bardeen, 1908–1991, American physicist and inventor, co-winner of the Nobel Prize in Physics in 1956 with his compatriots William Shockley (1910–1989) and Walter Houser Brattain (1902–1987) for their work on semiconductors and the discovery of the transistor effect at the Bell Laboratories.

The year 1948 brought a new invention in the world of microphones, the first dual-diaphragm microphone with dual directionality (see section 3.2.1.8.3 below). It was invented by G. Neumann GmbH[17] and his entire engineering team.

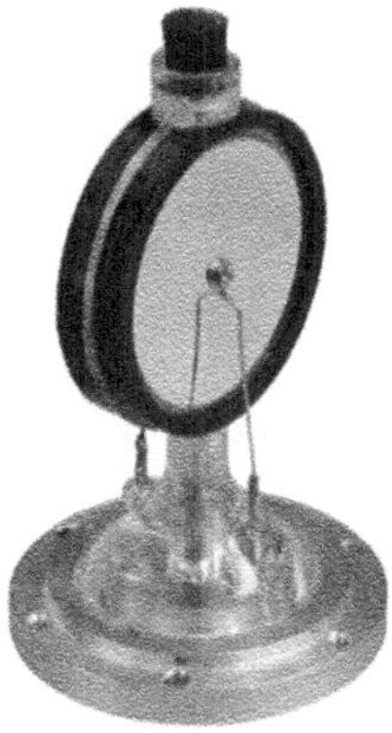

**Figure 3.15.** *The double membrane capsule, named M7, built by G. Neumann. For a color version of this figure, see www.iste.co.uk/reveillac/recording.zip*

During the next 20 years, the technologies of the various microphones did not cease to progress, the wireless microphone, the unidirectional microphone imvented by E. Seeler of the Shure company, the *electrostatic microphone* with *electret* (ECM, Electret Condenser Microphone), which is a condenser microphone that does not require a polarization charge, as the material that it is composed of has a permanent electrostatic charge. It was invented by G. Sessler and J. E. West in the laboratories of the Bell company.

The advent of transistors and semiconductors in general has lead the industry to use these new components, which have gradually replaced the vacuum tubes present in many microphones.

From 1983 onwards, electret microphones were often replaced by MEMS (MicroElectroMechanical System) technology, which makes them more reliable, more robust and smaller (a few millimeters).

17 Georg Neumann GmbH is a German company founded in 1928 in Berlin with the eponymous name of its founder. It is an emblematic manufacturer of microphones. Following its destruction by the Allies in 1943, the factory moved to Gefell to become Microtech Gefell during the reunification. At the same time, in 1949, Georg Neumann GmbH was recreated, still in Berlin.

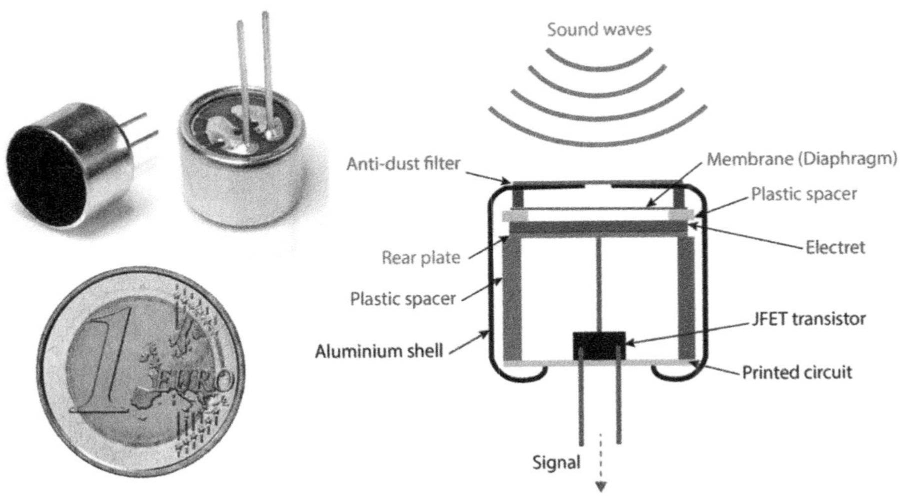


**Figure 3.16.** *The electret microphone and its principle. For a color version of this figure, see www.iste.co.uk/reveillac/recording.zip*

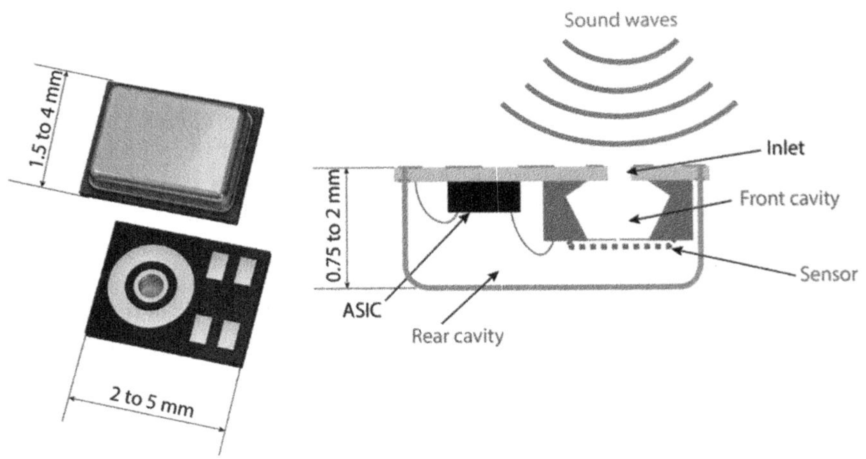


**Figure 3.17.** *The MEMS microphone and its principle. For a color version of this figure, see www.iste.co.uk/reveillac/recording.zip*

Figure 3.18 summarizes the various technologies associated with the history of the microphone in the form of a timeline.

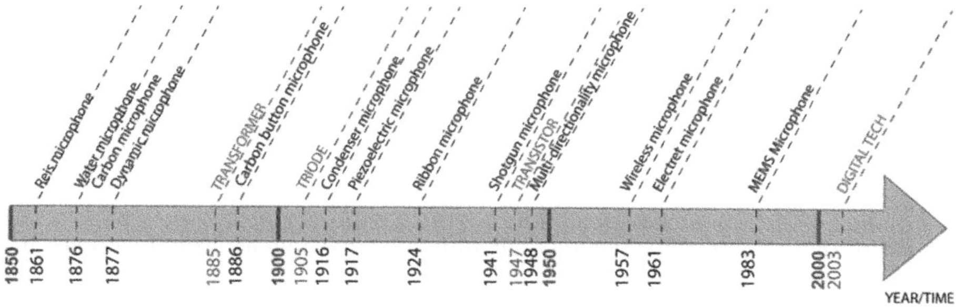


**Figure 3.18.** *Evolution of the microphone over time. For a color version of this figure, see www.iste.co.uk/reveillac/recording.zip*

From 1990 onwards, digital electronics, networks, the advent of the Internet and the Web once again revolutionized technologies, and audio systems were not left out. Digital audio developed and analog–digital converters appeared in the body of the microphones in 2013, making it possible for them to provide a digital signal at the output. The first *USB microphones*[18] came into existence in 2005.

## 3.2. The characteristics of a microphone

After this brief history, it seems necessary to me to define the characteristics of a microphone, which can be classified into two families, the first and main one being shared by all the models and the second one, more specific, attached to well-defined types.

### 3.2.1. *General characteristics*

#### 3.2.1.1. *Transducer technology*

The *transducer* or *capsule* may be based on very different technologies, which were discussed earlier in the history of microphones.

18 USB (Universal Serial Bus): a standard associated with a serial computer bus that makes it possible to connect various peripherals to a computer or to a device designed for this purpose. It is also capable of powering some of these devices. The standard has evolved over time from version 1.0 in 1996 (12 Mbit/s) to the official version 4.0 in 2019 (40 Gbit/s). There are several types of connectors: type A, type B, mini-connector, micro-connector and type C.

Here is a summary of the different types that are mainly found in studios today:

– dynamic moving coil microphone;

– dynamic ribbon microphone;

– condenser microphone.

All of these transducers integrate one of the two existing acoustic functions: the *pressure principle* or the *pressure gradient principle.*

In theory, in pressure microphones, the scalar pressure force acting on the outside of the diaphragm is considered so that the transducer response is independent of the angle of incidence of the sound wave, as long as the wavelength (generally up to a frequency of 1 kHz) is greater than the dimensions of the diaphragm. In fact, the membrane does not constitute an obstacle and does not disturb the propagation of the sound message. The inner side of the membrane is enclosed in a medium with a fixed pressure.

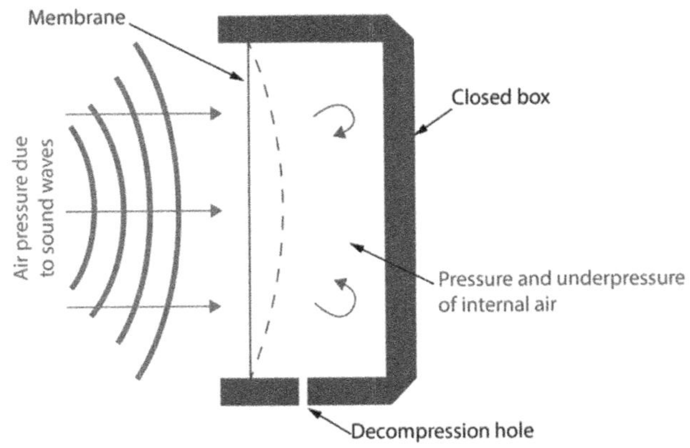


**Figure 3.19.** *Pressure transducer. For a color version of this figure, see www.iste.co.uk/reveillac/recording.zip*

When the length is equal to or less than the size of the membrane, the latter becomes an obstacle; the sound waves undergo interference with a *diffraction* phenomenon. They will be reflected and go back in the opposite direction (*reflected waves*), forming *standing waves* and an overpressure in the axis of the microphone, and this can go up to +6 dB. This phenomenon is taken into account by manufacturers who see a potential gain in clarity in the sound recording.

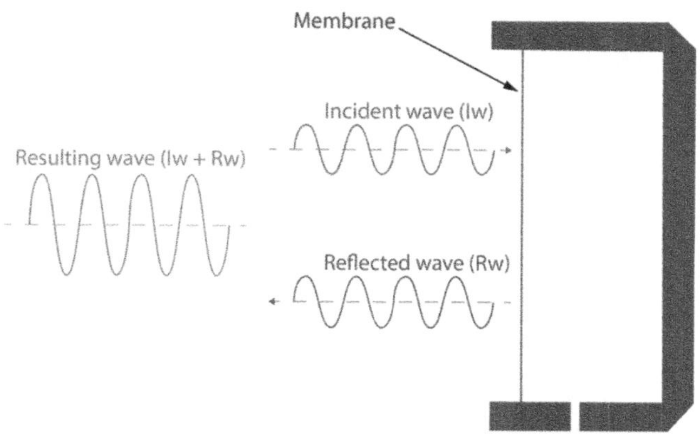


**Figure 3.20.** *Sum of incident and reflected waves. For a color version of this figure, see www.iste.co.uk/reveillac/recording.zip*

A second phenomenon, for the higher frequency waves, will appear when they hit the membrane, a phase shift, which will bring random wave cancellations and promote a significant drop in the output level.

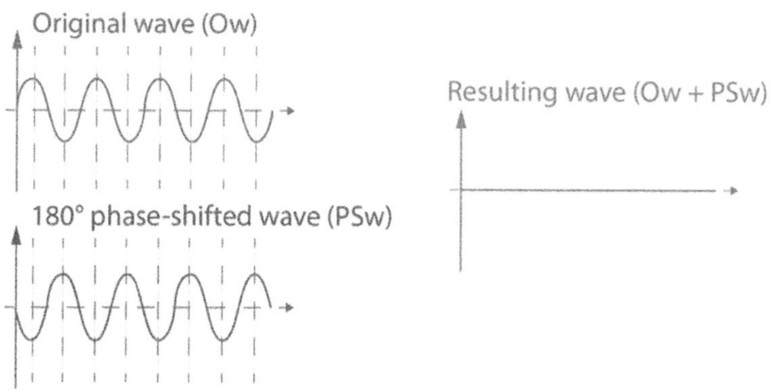


**Figure 3.21.** *A phase shift of 180° cancels the signal. For a color version of this figure, see www.iste.co.uk/reveillac/recording.zip*

In summary, we can say that pressure transducers are omnidirectional for frequencies below 1 kHz and increasingly directional up to 10 kHz and more.

For the pressure gradient, the active force is due to the difference in pressure acting on the front and back of the membrane; there is no closed insulating cavity behind the membrane. The sound vibrations strike the diaphragm from both sides.

The resulting signal is, therefore, a difference, called a *gradient*, between the signals arriving at the same time on each side. In this case, this force is dependent on the angle of incidence.

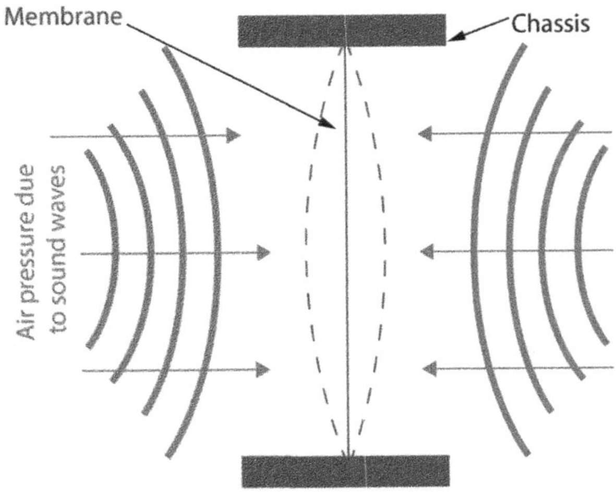


**Figure 3.22.** *Acoustic operation of a pressure gradient microphone. For a color version of this figure, see www.iste.co.uk/reveillac/recording.zip*

If the sound source is perpendicular to the axis of the transducer, at 90° or 270°, the signal arrives in phase on both sides of the membrane where it cancels out and the membrane remains stationary. Two opposite forces of the same intensity cancel each other out.

If the source is in the axis of the transducer, at 0° or 180°, the signal arrives alternately on the two sides of the membrane and the pressure difference between the two sides is maximum. It is, thus, in this case that the sensitivity is the most important.

The resulting force decreases with distance from the transducer axis until it is cancelled out when the transducer is in the perpendicular axis. This phenomenon provides a symmetrical directivity on each side of the diaphragm, which is called *bidirectional* or *figure-8 directivity* (see section 3.2.1.8.3).

### 3.2.1.2. *Noise level*

All microphones generate EIN (*Equivalent Input Noise*) themselves. This noise can be caused by diaphragm movement, thermal noise, noise from active components, etc. This noise is usually measured in sound pressure, Pa or dB SPL[19], or even in dBA[20].

Brownian motion is related to the random movement of particles in a fluid. In the case of microphones, the fluid involved is the ambient air. The thinner and lighter the microphone diaphragm, the more Brownian motion affects it. This is the case for ribbon and condenser microphones, which are more sensitive than dynamic moving-coil microphones. This noise is insignificant in most cases, although its presence is real.

*Johnson–Nyquist* noise, also called resistance noise or *thermal noise,* is the noise generated by the thermal agitation of electrons in an electrical conductor. It resembles *white noise* with a virtually continuous intensity over the entire audible frequency range. Depending on the output impedance, this noise varies; it is very low for a low impedance but is also sensitive to temperature. All conductors generate thermal noise, although it is often negligible.

In some microphones, the active circuits, such as impedance converters, filters, transformers or amplifiers, also generate noise due to the active electronic components that they are made up of and through which the audio signal from the transducer passes.

Noise affects the quality of the audio signal coming out of the microphone; so, care must be taken to maintain the correct *signal-to-noise ratio.*

Acceptable noise levels should be below 24 dBA. Between 11 and 15 dBA, the noise is hardly noticeable.

#### 3.2.1.3. *Signal-to-noise ratio*

This ratio, also called SNR, has an important role. It is expressed in dB and compares the signal power to the noise power. It must be as small as possible.

#### 3.2.1.4. *Output impedance*

The impedance measures the opposition of an electrical circuit to the passage of an alternating current. It is the equivalent of the resistance for a direct current. We can proceed to a generalization of Ohm's law by applying it to the alternating current, that is, Z=U/I with U, the voltage and I the intensity of sinusoidal forms. This term was coined by the British physicist Oliver Heaviside in 1886.

19 See Appendix 1.

20 See Appendix 1.

Impedance is measured in ohms like resistance. It is most applicable to microphones that supply signals in the form of alternating currents.

It is measured at the output terminals of the microphone when it is connected to a preamplifier system. The impedance of the preamplifier must, therefore, also be taken into account.

There are, in fact, two distinct impedances: the output impedance, which is inherent in the microphone, and the load impedance, which is the input impedance of the device to which the microphone is connected. It is considered that the output impedance of the microphone should be on average 10 times lower than the load impedance.

In order for the microphone signal to be effectively transmitted through the connection, the load impedance must always be greater than the output impedance of the microphone.

The impedance of the microphones can be summarized as follows:

– low impedance: <600 Ω;

– average impedance: 600 Ω to 10 kΩ;

– high impedance: >10 kΩ.

Professional microphones are always low impedance and range from 50 to 600 Ω.

The low impedance of professional microphones is there to make it possible for them to have significant cable lengths without affecting the quality of the transmitted audio signal before pre-amplification or other processing.

When a microphone is connected to the microphone input of a peripheral device (generally, the connection is of the XLR type[21]), preamplifier, mixing console, effects rack, etc., the expected input level is generally between 1 and 10 mV (–60 dBV and –40 dBV), with an input impedance of between 1.5 kΩ and 15 kΩ.

In the case of condenser microphones, the very high impedance of the transducer necessarily leads to the use of an *impedance converter*, based on the tube or transistor technology.

21 See Appendix 2.

#### 3.2.1.5. *Sensitivity*

It measures the efficiency of the microphone, that is, its ability to transform the acoustic energy of a sound into electrical energy. If we know the acoustic pressure to which the membrane of a microphone is subjected and we measure its output voltage, we can determine its sensitivity.

To carry out the measurement, the microphone is subjected to an acoustic pressure of 1 Pa[22] (pascal) at the input, that is, 94 dB SPL[23] supplied by an audio signal at 1 kHz, and then its output voltage is measured in millivolts. It can also be expressed in dBV[24].

Low-sensitivity microphones are used for strong acoustic signals, whereas high-sensitivity microphones are used for weak sounds.

In the specifications for microphones, manufacturers give an impedance called the nominal impedance, which is an average of the output impedance for frequencies ranging from 20 Hz to 20 kHz.

Passive microphones are generally much less sensitive than their active counterparts, which have an internal amplification device.

Sensitivities can be classified according to technologies:

– ribbon microphones: 0.5–8 mV/Pa[25];

– passive dynamic moving coil microphones: 1–8 mV/Pa;

– active ribbon microphone: 6–32 mv/Pa;

– active condenser microphones: 8–32 mv/Pa.

The sensitivity of a microphone is fixed, although its output volume can be changed. It can also vary with directionality for directional microphones. The thinness of the diaphragm is not the main criterion for sensitivity; it has more to do with the quality and the rendering of the audio signal.

The sensitivity is rather dependent on the strong reactivity of the diaphragm to the sound signals, to the type of transducers, capacitors, ribbons, moving coils, etc., and even to the amplification systems.

---

22 See Appendix 1.

23 See Appendix 1.

24 See Appendix 1.

25 See Appendix 1.

### 3.2.1.6. *Frequency response*

The frequency response of a microphone is one of the key parameters that defines the sound that will be obtained with a particular microphone. It shows the output level as a function of frequency. It is presented in the form of a graph, showing the frequencies in hertz on the x-axis and the relative sensitivity in dB on the y-axis.

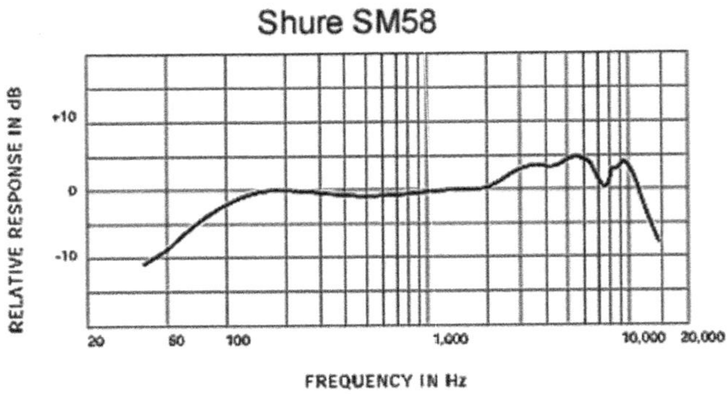


**Figure 3.23.** *The frequency response curve of a Shure SM58 Microphone (source: www.shure.com)*

Humans do not hear sound frequencies linearly, but logarithmically[26], which explains the graduation of the x-axis scale between 20 and 20,000 Hz, the frequency range of the sound spectrum audible by the human ear. Below 0 dB, the microphone is less sensitive; above 0 dB, its sensitivity increases.

With this graph, you can quickly see the frequency areas where the microphone is more or less effective.

This curve does not allow you to easily imagine the sound color provided by the microphone; only a test under real conditions will make it possible for you to hear it.

This characteristic is one of the most important in defining a microphone. It is related to many factors that are internal and external to the microphone. The following are some of the most common ones that can be cited:

– the size and shape of the capsule;

– the size of the diaphragm;

– the shape of the diaphragm;

26 See Appendix 1.

– the weight of the diaphragm;

– capsule orientation;

– the distance from the sound source;

– output and load impedance;

– the grid and the cushioning material around the capsule;

– the space around the capsule;

– the shape of the microphone;

– the nature and quality of the impedance converter;

– the resonant frequency of the microphone.

#### 3.2.1.7. *Maximum allowable sound pressure*

This pressure is measured in dB SPL[27]. It indicates the maximum level that can be supported by a membrane before saturation occurs or the signal is degraded.

It depends on the technology of the microphone (condenser, electret, dynamic) and the size of its diaphragm.

#### 3.2.1.8. *Directivity*

For a microphone, the directivity represents its sensitivity as a function of the direction from which the sound comes. It is represented by a polar diagram.

This type of diagram is made up of concentric circles, whose center corresponds to the microphone to be studied. Each circle represents the surrounding acoustic space over 360°. 0° represents the front of the microphone and 180° the back. Each of the circles is graduated in dB to specify a sound attenuation value.

Superimposed on this reference frame are one or more curves attached to a frequency, which characterize the sensitivity of the microphone in space as a function of the angle between the source and the front of the microphone.

The nature of the directionality influences the so-called *proximity effect*, which is noticeable in areas close to the transducer where a phase shift of the sound waves develops, leading to a substantial amplification of low frequencies.

27 See Appendix 1.

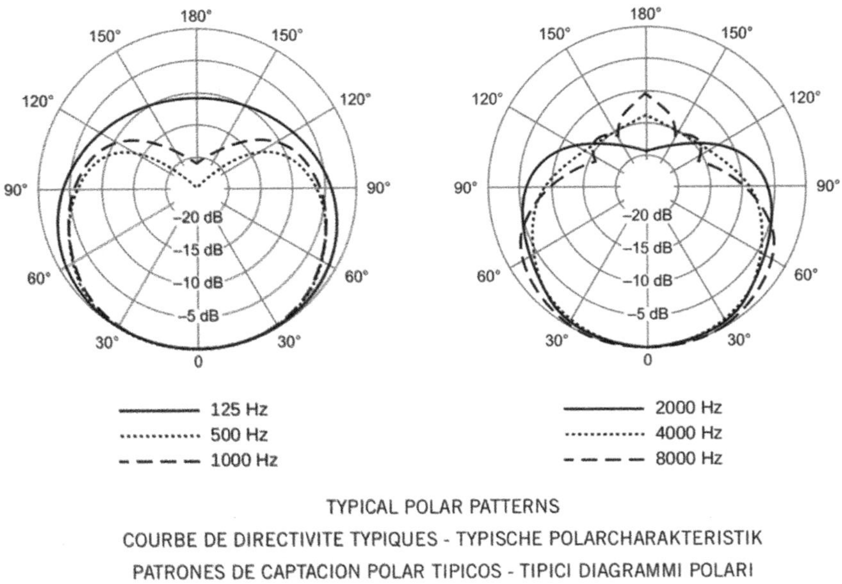


**Figure 3.24.** *Directivity diagrams of a Shure SM58SE microphone (source: www.shure.com)*

Therefore, these curves indicate the directionality, which can be divided into several types:

– *omnidirectional*;

– *hemispheric*;

– bidirectional;

– *cardioid*;

– *supercardioid*;

– *hypercardioid*;

– *subcardioid*;

– canon.

### 3.2.1.8.1. Omnidirectional

A microphone with an omnidirectional pickup pattern should theoretically be equally sensitive in all directions, regardless of the frequency of the sound message being captured. It operates on the pressure principle (see section 3.2.1.1).

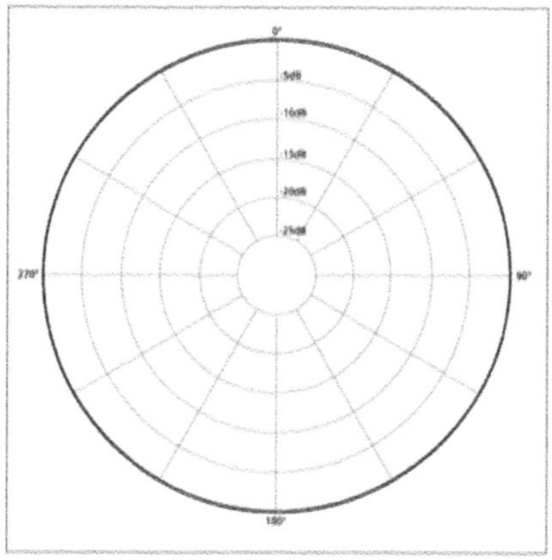

**Figure 3.25.** *The directivity diagram for a perfect omnidirectional microphone*

NOTE.– This is true for low frequencies, but not for higher frequencies where most transducers tend to be unidirectional. The wavelength of high frequencies is very short, so the nature of the microphone body affects the directionality.

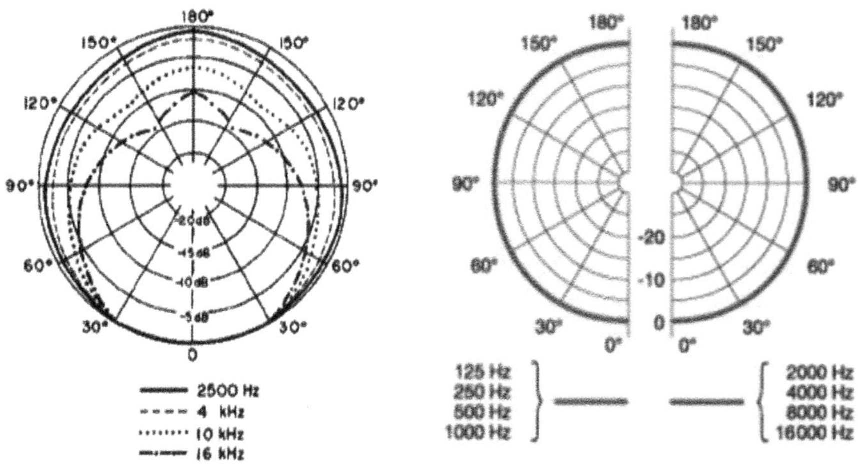


**Figure 3.26.** *The directivity diagram for the Shure SM80-LC omnidirectional microphones, left, and AKG LC617-MD, right (source: www.shure.com and www.akg.com). For a color version of this figure, see www.iste.co.uk/reveillac/recording.zip*

Omnidirectional microphones have an excellent grip on natural sound phenomena with very little coloration, and are very resistant to vocal plosives[28] and wind effects.

### 3.2.1.8.2. Hemispheric

This directionality is reserved for microphones, also called PZM (Pressure Zone Modification). They are often not well known, yet they provide an interesting solution in many situations. Their invention is due to the engineer E. M. Long[29].

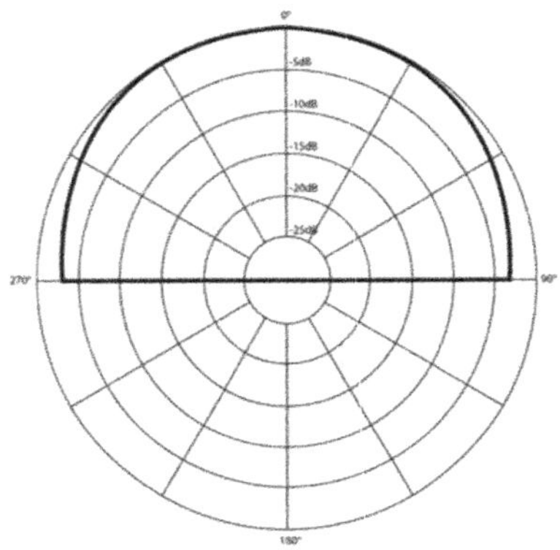


**Figure 3.27.** *The directivity diagram for a perfect hemispherical microphone*

Its shape and appearance seem quite unusual; it is flat and works placed against a flat surface of a fairly large size, about 1 $m^2$.

This microphone operates on the pressure or pressure gradient principle. It is usually designed around an omnidirectional or cardioid transducer, or even a supercardioid or hypercardioid transducer (see sections 3.2.1.8.4–3.2.1.8.6).

Its position against a flat surface eliminates sound reflections that could enter from behind. It works best if the surface against which it is pressed is very large.

---

28 Vocal plosives: When sounds such as "p" and "b" are uttered, a strong gust of air is expelled through the mouth. On a recording, these bursts of air hit the microphone diaphragm, creating a low, thumping sound known as a vocal plosive.

29 Edward M. Long, 1932–2016, American engineer, worked on studio monitoring systems, loudspeaker crossovers and many other audio systems. He was a consultant for the American company Crown Audio, where he developed the PZM microphone.

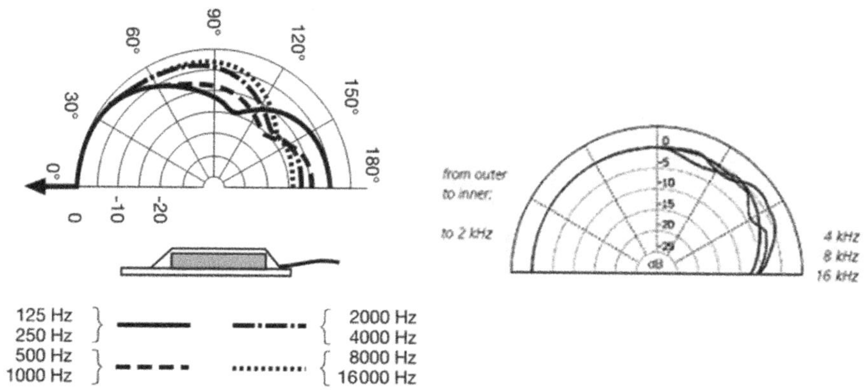


**Figure 3.28.** *The directivity diagram for the AKG C547BL hemispherical microphone, left, and the Schoeps BLM, right (source: www.akg.com and www.dcaudiovisuel.com)*

### 3.2.1.8.3. Bidirectional

Bidirectional directional microphones, also called figure-of-eight microphones, have the same sensitivity to the front and back. When you look at the diagram, you can see a figure-of-eight curve. It is based on the pressure gradient principle (see section 3.2.1.1). Ribbon microphones are almost always bidirectional.

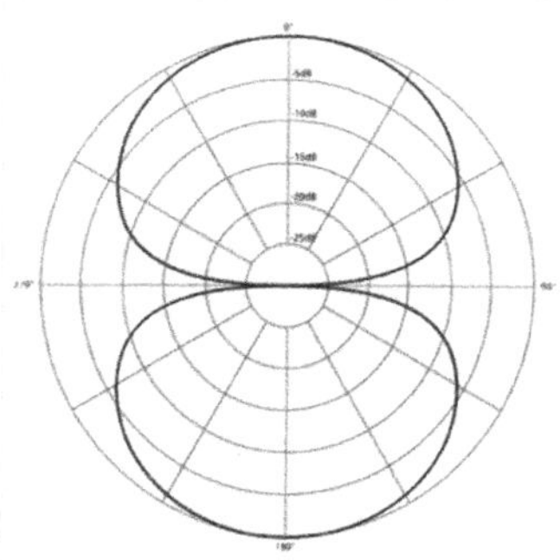

**Figure 3.29.** *The directivity diagram for a perfect bidirectional microphone*

Some microphones have a bidirectional pickup pattern because they consist of two back-to-back cardioid capsules of equal sensitivity and opposite polarity.

It is physically impossible to achieve true front-to-back directivity in a microphone; around the lateral axes, a small portion of the sound signal will still be picked up as shown in the polar diagram.

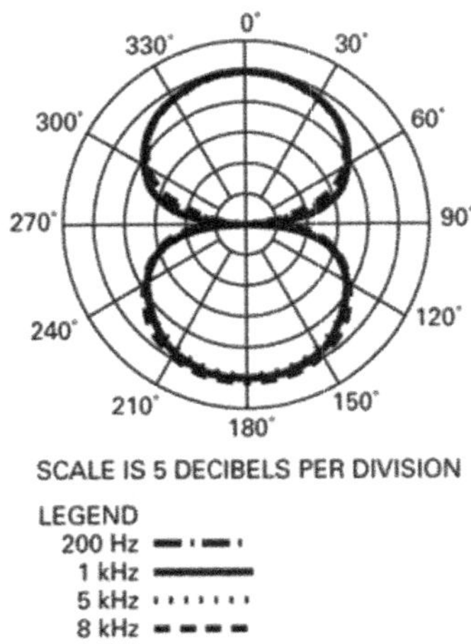


**Figure 3.30.** *The directivity diagram for the Audio-Technica AT4080 bidirectional microphone (source: www.audio-technica.com)*

As with other microphones, a bidirectional microphone is sensitive to vocal plosives and *feedback*[30], especially laterally. Its directivity becomes more important at higher sound frequencies.

### 3.2.1.8.4. Cardioid

In order to accept this directivity, microphones must have a heart-shaped directivity pattern, giving them their name. Often, they are also associated with the term unidirectional and work with the pressure gradient principle (see section 3.2.1.1).

They are widely used and are mainly sensitive to sounds coming from the front. The attenuation reaches on average –6 dB at 90° and 270°.

30 Feedback occurs when a transmitter and a receiver, for example, the speaker and microphone of an audio system, are placed close to each other. The sound emitted by the transmitter is picked up by the receiver, which transmits the amplified sound to the transmitter. This signal is again picked up by the transmitter and re-amplified and so on. The result is the creation of a very powerful whistling sound.

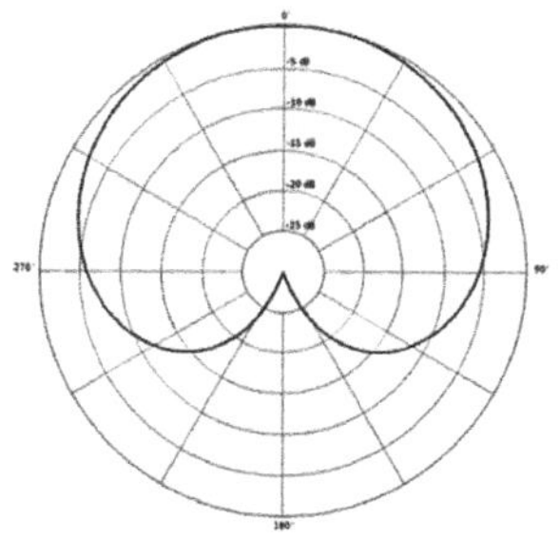

**Figure 3.31.** *The directivity diagram for a perfect cardioid microphone*

One of their peculiarities is the change in color of the sound signal depending on the lateral position of the source, which is a significant problem. One of their strengths is that their forward-directed polarity, 0 dB at 180°, makes them less sensitive to feedback effects on stage, since the maximum sound rejection is to the rear. They are less sensitive than a conventional bidirectional microphone. Beware of vocal plosives, for which a cardioid transducer is sensitive.

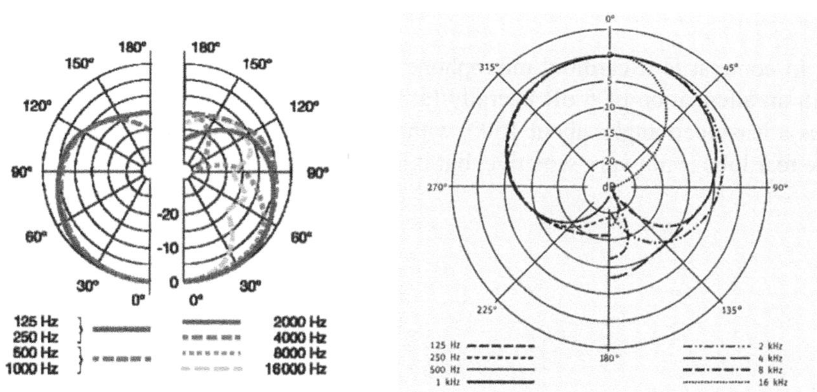


**Figure 3.32.** *The directivity diagram for the Shure D5C cardioid microphone on the left and the Neumann TLM103 on the right (source: www.shure.com, www.neumann.com). For a color version of this figure, see www.iste.co.uk/reveillac/recording.zip*

These are the most common microphones on the market and the most used for voices, whether in the studio, in an interview or in a *show*[31]. They exist in many configurations, small or large diaphragm, dynamic or condenser technology.

31 Lavalier microphones are also called lapel microphones. They are small and attach to your shirt, collar or tie. They make it possible to record hands-free.

### 3.2.1.8.5. Supercardioid

This pattern is related to the cardioid polar pattern, although it is more directional. It has two null points at 127° and 233° and a small rear sensitivity lobe, but like the cardioid pattern, the maximum receptivity is towards the front where it becomes very directional.

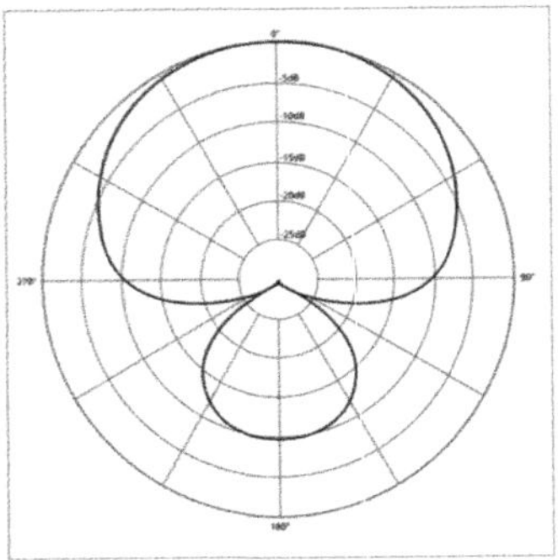

**Figure 3.33.** *The directivity diagram for a perfect supercardioid microphone*

In contrast to a cardioid microphone, which has a pickup angle of about 180°, with an attenuation of 6 dB laterally (at 90° from the main axis), the supercardioid uses a less open angle, about 150°, with a pickup that drops by 3 dB on each side. The rear lobe is not very sensitive, but it is not negligible.

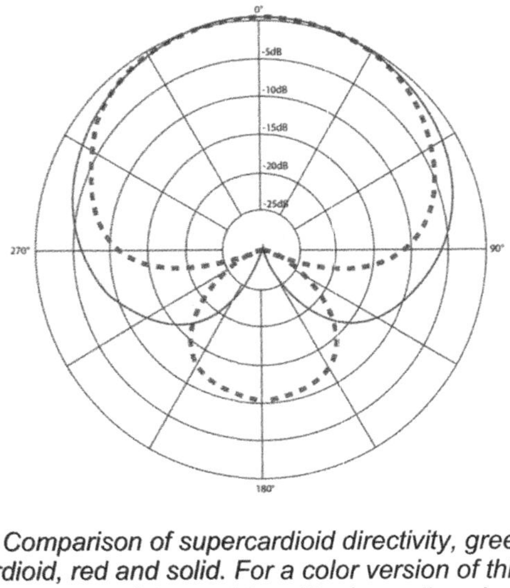


**Figure 3.34.** *Comparison of supercardioid directivity, green and dotted, and cardioid, red and solid. For a color version of this figure, see www.iste.co.uk/reveillac/recording.zip*

Supercardioid transducers operate on the pressure gradient principle, and are therefore sensitive to the proximity effect.

This type of microphone is often chosen for use in movies and noisy environments. As mentioned above, the back lobe exists despite an average attenuation of 10 dB, depending on the frequency, and should therefore be taken into consideration. The supercardioid transducer should be chosen in cases where it is desired to isolate individual sound sources in difficult situations.

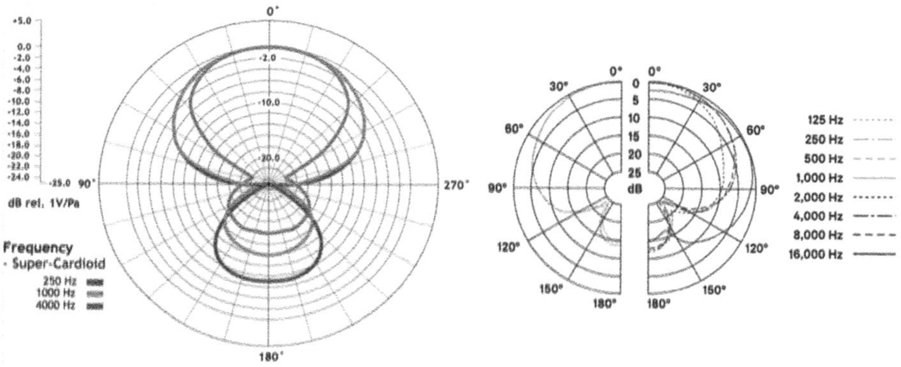


**Figure 3.35.** *The directivity diagram for the Rode NTG-2 supercardioid microphone, left, and the Sennheiser E614, right (source: www.rode.com, www.sennheiser.com). For a color version of this figure, see www.iste.co.uk/reveillac/recording.zip*

Beware of wind noise and vocal plosives, for which a supercardioid microphone is very sensitive.

### 3.2.1.8.6. Hypercardioid

Although it has the same suffix, the hypercardioid directional pattern is different from its cardoid, supercardioid and subcardioid sisters. Hypercardioid transducers have a very high directivity, are very sensitive to axial sounds, have two null points at 110° and 250° and a rear lobe (–6 dB at 180°).

The hyperdirectivity of these microphones is reminiscent of the combination of bidirectional and cardioid polar patterns. A lateral attenuation of 12 dB is present at 90° and 270°. On the bangs of the acceptance angle, which is 150°, a loss of 3 dB is observed.

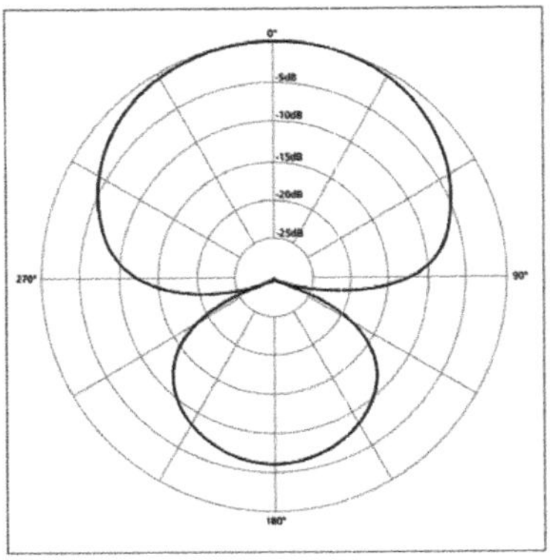

**Figure 3.36.** *The directivity diagram for a perfect hypercardioid microphone*

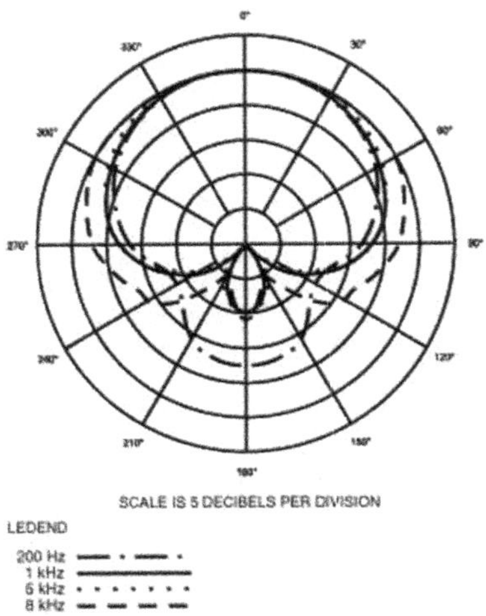


**Figure 3.37.** *The directivity diagram for the Audio Technica AT4053b supercardioid microphones (source: www.rode.com, www.audiotechnica.com)*

This type of transducer is sensitive to plosive vocals and has a proximity effect. It is very directional in the high frequency ranges with a slight loss in the low end where it approaches a cardioid pattern.

This type of microphone is often fitted with an interference tube, allowing it to evolve towards a shotgun directivity (see section 3.2.1.8.8).

### 3.2.1.8.7. Subcardioid

With this pickup pattern, a microphone can easily pick up a signal from the front, as well as from the side and the rear with a slightly lower amplitude. It is a compromise between an omnidirectional and a cardioid pickup pattern.

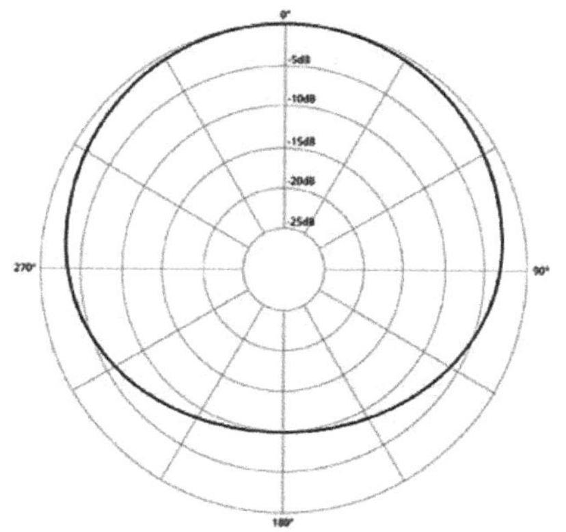

**Figure 3.38.** *The directivity diagram for a perfect subcardioid microphone*

Its lowest sensitivity is at 180°, where it drops by 10 dB. At 90° and 270°, it drops by 3 dB.

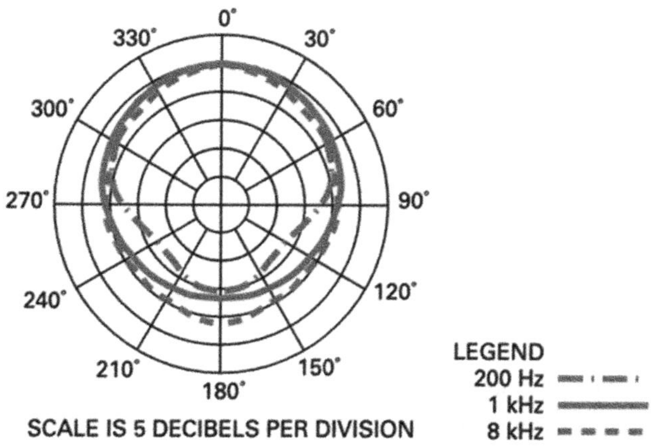


**Figure 3.39.** *The directivity diagram for the Audio Technica AT808G subcardioid microphone (source: www.audiotechnica.com). For a color version of this figure, see www.iste.co.uk/reveillac/recording.zip*

These microphones have a transducer that operates with a pressure gradient. At the back of the diaphragm, an acoustic labyrinth dampens and delays the sound, which gives this type of transducer a pattern close to omnidirectional directivity in the 180° axis.

Its coloration is generally weak, and the sound message remains very natural. Its absence of zero point makes it a transducer sensitive to sounds coming from all directions, little impacted by proximity effects. It is quite resistant to vocal plosives.

### 3.2.1.8.8. Gun or shotgun

Microphones with this directionality are widely used in the audio-visual, film and sound recording industries. They are very directional, and as we have seen above, they use very long interference tubes that make them easy to recognize. Although their sensitivity is mainly directed along the frontal axis, they have small lateral and rear sensitivity lobes.

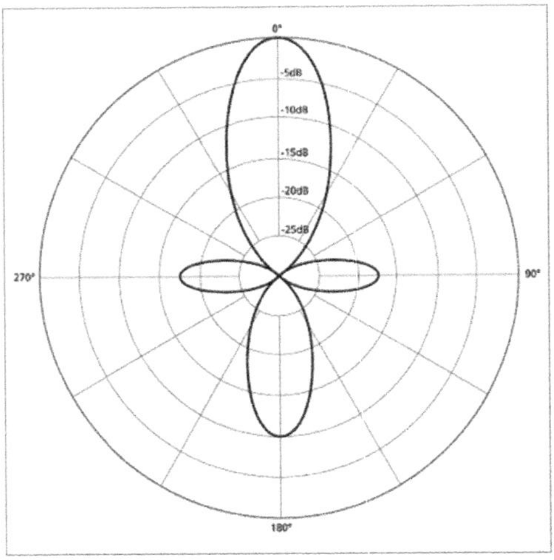


**Figure 3.40.** *The directivity diagram for a perfect shotgun microphone*

In practice, the side lobes are not very troublesome. The operation is based on the pressure gradient principle (see section 3.2.1.1). Generally, the rear lobe has a sensitivity of −10 dB compared to the front lobe. Internally, the transducer is supercardioid or hypercardioid.

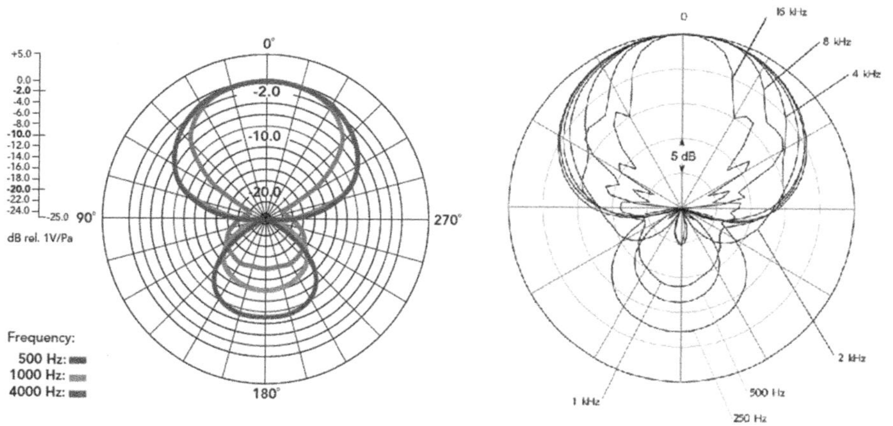


**Figure 3.41.** *Directivity diagram for the Rode NTG-1 shotgun microphones, left, and DPA 4017B (source: www.rode.com, www.dpamicrophones.com). For a color version of this figure, see www.iste.co.uk/reveillac/recording.zip*

## 3.2.2. *Specific characteristics*

### 3.2.2.1. *Membrane width*

When a sound signal is picked up, the pressure variation that the membrane undergoes, due to the agitation of the air molecules around their equilibrium position, is faster when the frequency is higher, that is, for higher pitched sounds. It is the succession of pressure and vacuum phases that sets the microphone membrane in motion so that it can transform them into an alternating electrical signal.

Depending on its surface and, therefore, indirectly on its weight, a large membrane has a high inertia compared to a smaller one.

At higher frequencies, the reciprocating action of the diaphragm, which is similar to that of a piston, makes the microphone less sensitive to high frequencies. Moreover, depending on the size, a microphone can also be subject to diffraction, which makes it an obstacle in the middle of the sound field, allowing the shortest, and therefore highest, sound waves to collide with its structure and be reflected, without even benefiting from the transduction generated by the membrane.

Small-diaphragm microphones are less sensitive because the diaphragm is less agile, so they will often require significant amplification, which will often result in significant background noise, unlike their large-diaphragm counterparts.

For a condenser microphone, a small diaphragm is considered to have a diameter of 12.7 mm (1/2") or less, whereas large diaphragms are around 2.54 cm (1").

A sound wave hitting a membrane generates a power $P$ (in W) proportional to the surface area $S$ (in $m^2$) and the square of the acoustic pressure $p$ (in Pa)

$$P = S.\frac{p^2}{p_0 c}$$

with $c$, the speed of sound, 343 m/s and $p_0$ the density of air, 1.2 kg/$m^3$.

Example:

Let's say that a perpendicular sound signal with a pressure of 0.8 Pa reaches a membrane of 22 mm diameter.

We obtain:

$$S = \pi R^2 = \pi \times \frac{0.022}{2} = 3.4557 \times 10^{-3} m^2$$

hence:

$$P = 3.4557 \times 10^{-3} \times \frac{0.8}{1.2 \times 343} = 0.67 \mu W$$

### 3.2.2.2. *Amplification type*

Condenser microphones include an electronic part to amplify the signal coming from the capsule. Two technologies are available: electronic tube and FET[32] transistor.

Both are available on the market, the first is appreciated for its coloring – warm, rich and soft – at the expense of the precision of the signal. The second, which is the most common, combines more coolness, with great accuracy and brilliance in the rendering of the sound message.

Both systems are impedance converters and amplifiers[33]. The tube system is fragile by design, while its competitor is very robust.

---

32 The FET (Field Effect Transistor) makes it possible to control, by means of an electric field, the conductivity of a channel within a semiconductor material.

33 See Appendix 4.

Tube microphones often require a specific external power source, whereas FET microphones only need *phantom* power or a DC power source, such as a battery or USB port[34].

NOTE.– Phantom power is a device that makes it possible for you to connect microphones that need power with ordinary microphone cables. It eliminates the need for a separate power supply unit, batteries or special cables. It provides a DC voltage of 48 V or 12 V (DIN 45596 standard) positive to ground.

**Figure 3.42.** *Two power supplies, the Neumann M149A model specifically for the Neumann M147 tube microphone and a conventional 48V phantom power supply*

## 3.3. Microphone families

After the history and description of the main characteristics of microphones and more generally, of their transducers or capsules, I will present some models of microphones that are more or less recent, but widely used, by going back over the technology they embody. We will look at moving coil dynamic microphones, ribbon microphones, condenser microphones and finally, electret microphones.

The set of microphones presented in the following paragraphs is far from being exhaustive. It is rather the fruit of my modest experience in relation to the use that I could have in conditions as diverse as live, studio or interview.

There are many others. A microphone is a bit like a musical instrument; you can only really get an idea of it after trying it out, taking into account the environment and the conditions in which it is used.

34 USB (Universal Serial Bus), a standard for a computer bus, dating from January 1996, capable of delivering a voltage of 5V DC.

### 3.3.1. *Microphone and transformer*

Before I start presenting some of the microphones of each family, I would like to introduce you to an important and often indispensable element, which I have not mentioned – the transformer. It is generally present in all professional microphones.

The transformers, invented by the Hungarians M. Déry, K. Ziperowsky and O. Bláthy between 1885–1886, made their appearance within the dynamic microphones at the beginning of the 1920s.

In the 1930s, high-powered magnets further improved the fidelity of dynamic microphones, which were widely used at that time.

The appearance of neodymium magnets[35] in the 1980s made it possible to create more intense magnetic fields, which further improved the quality.

In 1931, E.C. Wente and A.L. Thuras further improved the quality of the moving coil devices with new materials and a more advanced manufacturing technology.

The role of the transformer is to modify the voltage, the current and the impedance by connecting, using electromagnetic induction, two circuits called primary winding and secondary winding.

The primary winding is the input of the transformer and the secondary is the output. A magnetic core connects them, as shown in Figure 3.41.

There is no electrical connection between the two windings, which are totally independent.

Transformers can be of several types depending on certain physical parameters related to the windings, such as the number of turns, the diameter of the wire, its resistivity, etc. There are step-up or step-down voltage transformers or insulation transformers.

35 Neodymium is a chemical element with an atomic number of 60 and has the appearance of a grayish metal. When used in an alloy with iron, it makes it possible to create powerful but fragile magnets.

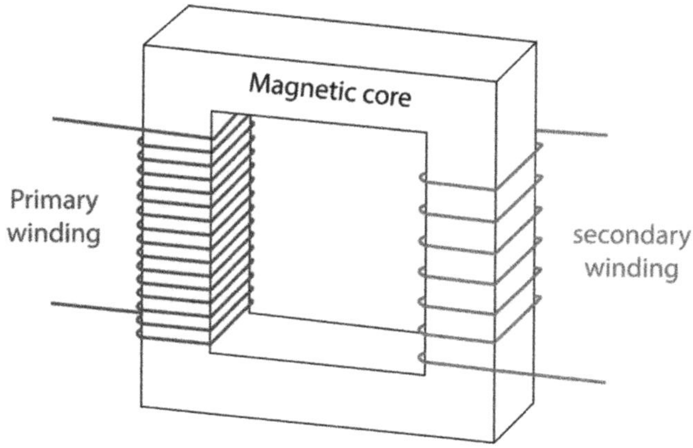


**Figure 3.43.** *A transformer with its core and two windings. For a color version of this figure, see www.iste.co.uk/reveillac/recording.zip*

In a microphone, the transformer can also act to filter a DC voltage, which could possibly damage internal components, and to balance and adapt the output. Its presence can also modify the coloring of the output audio signal by adding distortion due to the imperfection of its constituent elements (windings, nature of the conductive wires, core material, etc.).

**Figure 3.44.** *A typical microphone transformer (14 x 10 x 12 mm). For a color version of this figure, see www.iste.co.uk/reveillac/recording.zip*

Many of the professional microphones on the market today have a transformer that serves, among other things, to balance the signal and match the output impedance.

NOTE.– The transformer is not an essential element and some microphones do not have one.

## 3.3.2. *Dynamic moving coil microphones*

Very common in the professional world, dynamic microphones have become increasingly sophisticated over time, with characteristics that perfectly correspond to live performances and recording instruments in the studio where the output level is important, such as guitar or bass amps, drums, brass instruments, etc.

This type of microphone is rather forgiving and will show up fewer faults than a condenser microphone (which requires a power supply). They usually carry a transformer.

### 3.3.2.1. *Some dynamic moving coil microphones*

In Table 3.1, I have only included microphones that are still readily available on the market, both new and used. These microphones are generally not very fragile and are moderately priced, although some can reach more prohibitive prices.

<table>
<tr><th>Brand</th><th>Model</th><th>Frequency response</th><th>Sensitivity</th><th>Directivity</th><th>Output impedance</th></tr>
<tr><td>AKG</td><td>D5</td><td>70 Hz to 20 kHz</td><td>2.6 mV/Pa</td><td>Supercardioid</td><td>600 Ω</td></tr>
<tr><td colspan="3">Designed for voice and choir, although it can be versatile. It delivers a significant sound level and its polarity ensures no feedback. It remains clear, precise and very linear in the midrange. Due to its design, its robustness is unparalleled.</td><td colspan="3">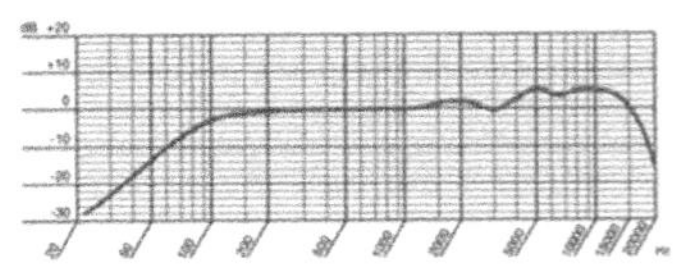</td></tr>
<tr><td>AKG</td><td>D112</td><td>20 Hz to 17 kHz</td><td>1.8 mV/Pa</td><td>Cardioid</td><td>210 Ω</td></tr>
<tr><td colspan="3">Specialized in recording sound instruments working in low frequencies such as bass drum, tuba, double bass, timpani, etc., or the dubbing of bass amps. Lots of roundness and warmth with a robust design.</td><td colspan="3">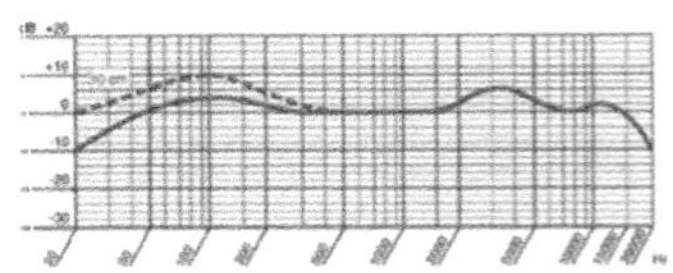</td></tr>
<tr><td>Beyerdynamic</td><td>M201TG</td><td>40 Hz to 18 kHz</td><td>1.2 mV/Pa</td><td>Hypercardioid</td><td>200 Ω</td></tr>
<tr><td colspan="3">A very natural, uncolored sound with great sensitivity. It is an all-purpose microphone, interesting for its small size. It is used for guitar amp pickup, snare drum or hi-hat pedal, brass instruments and even vocals.</td><td colspan="3">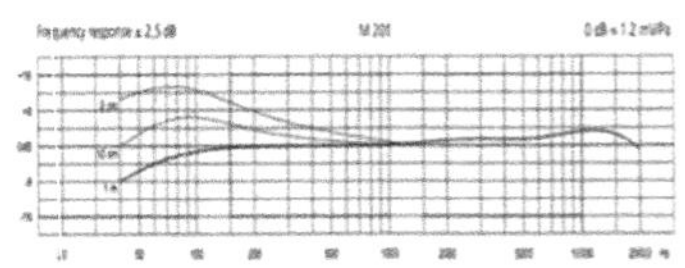</td></tr>
</table>

| Brand | Model | Frequency response | Sensitivity | Directivity | Output impedance |
|---|---|---|---|---|---|
| Beyerdynamic | M88TG | 30 Hz to 20 kHz | 2.9 mv/Pa | Hypercardioid | 200 Ω |
| A robust product that has been around since the 1960s. It is versatile with a warm yet natural sound. Frequently used on the bass drum, bass amps, some brass instruments for its ability to capture the low end of the spectrum. | | | 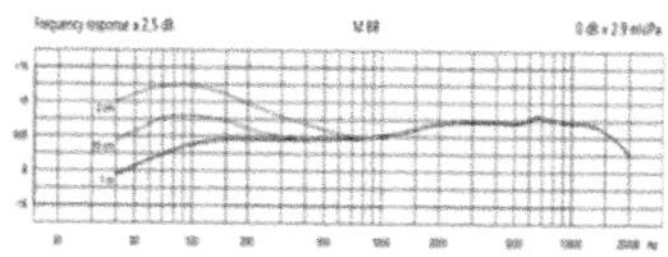 | | |
| Sennheiser | E906 | 40 Hz to 18 kHz | 2.2 mV/Pa | Supercardioid | 350 Ω |
| Mainly dedicated to amp patching, but can also be used on percussion or brass instruments. It supports high acoustic pressures. | | | 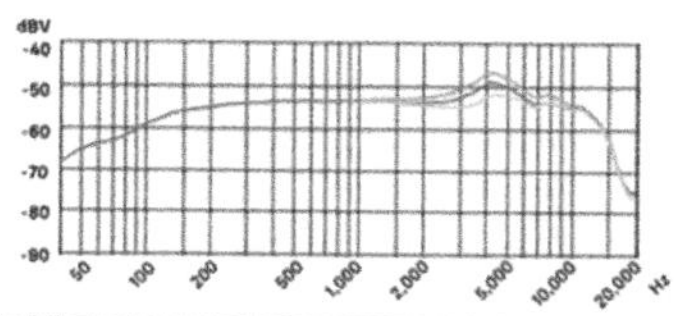 | | |
| Sennheiser | MD441 | 30 Hz to 20 kHz | 1.8 mV/Pa | Supercardioid | 200 Ω |
| Its excellent sensitivity and high linearity in the bass and midrange make it a high-end microphone with a very neutral sound. It includes a 5-position equalizer for making adjustments. | | | 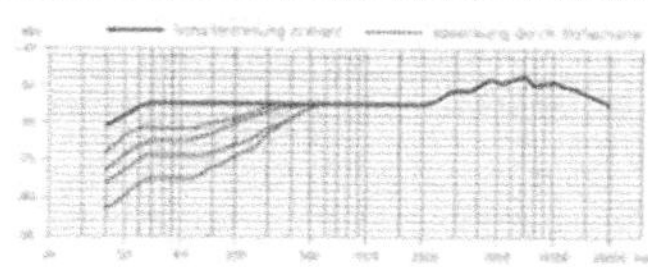 | | |
| Shure | SM57 | 40 Hz to 15 kHz | 1.6 mV/Pa | Cardioid | 310 Ω |
| Indispensable. A reference in the field of dynamic microphones. Very robust. Suitable for vocals, brass, drums, amps. Supports high acoustic pressure. | | | 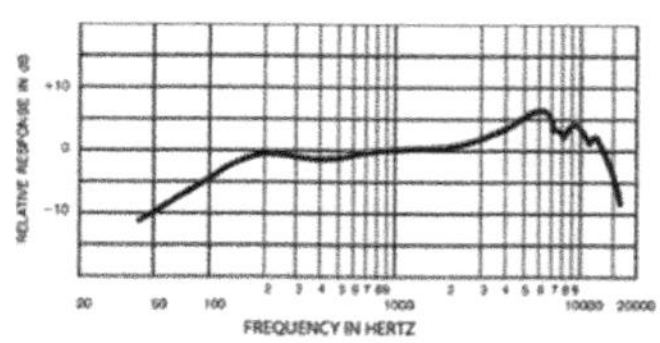 | | |
| Shure | SM58 | 50 Hz to 15 kHz | 1.85 mV/Pa | Cardioid | 300 Ω |
| Like the SM57, this microphone is a star performer, the most widely used microphone in the world for both spoken and sung voices. It is completely suitable for live use and its grille minimizes wind and breath noise. Its directional pattern easily isolates unwanted background noise and it is extremely robust. | | | 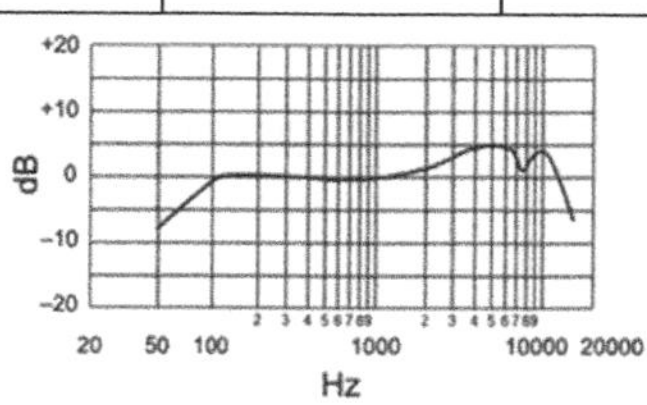 | | |

| Brand | Model | Frequency response | Sensitivity | Directivity | Output impedance |
|---|---|---|---|---|---|
| Shure | SM7B | 50 Hz to 20 kHz | 1.12 mV/Pa | Cardioid | 150 Ω |
| A rugged mic with excellent performance on vocals and guitar amp pickups. It has a built-in filter to control bass and midrange. Its rejection of sounds coming from the rear makes it an easily positionable mic. | | | 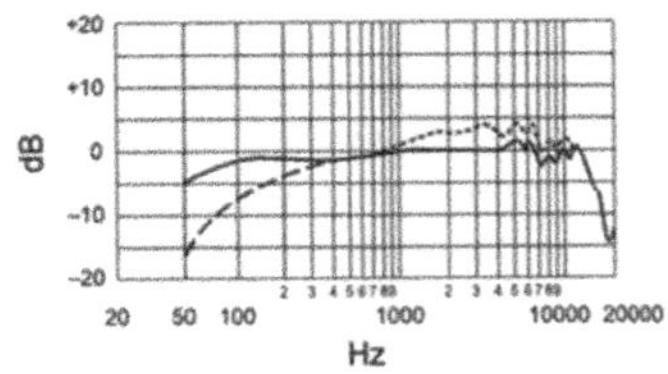 | | |
| Telefunken | M80 | 50 Hz to 18 kHz | 1.54 mV/Pa | Supercardioid | 325 Ω |
| A very good microphone for voices, which it leaves very natural and detailed. The presence of a peak between 5 and 10 kHz makes it a very sharp-sounding microphone in the style of today's recordings. Some use it to pick up a hi-hat pedal or a snare drum. | | | 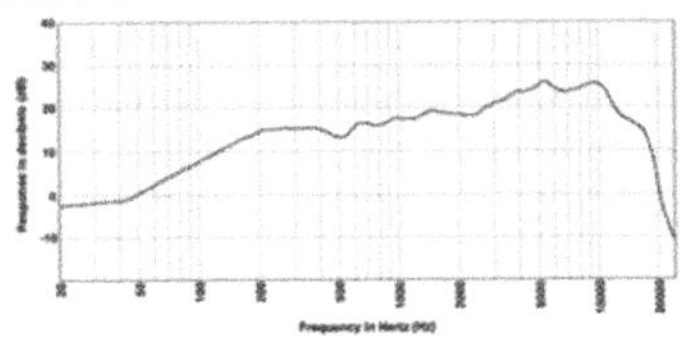 | | |

**Table 3.1.** *Some models of dynamic moving coil microphones with their main characteristics*[36]. *For a color version of this table, see www.iste.co.uk/reveillac/recording.zip*

### 3.3.3. ***Ribbon microphones***

Ribbon microphones fall into the category of dynamic microphones, but they can also be operated with phantom power. Their directionality is often bidirectional and they are almost always used in a studio with a perfectly controlled sound environment.

In terms of sound rendering, they are precise, warm and very round.

#### 3.3.3.1. *Some ribbon microphones*

Table 3.2 shows some of the ribbon microphones available on the market. These microphones are often expensive and very fragile; improper storage, shock, use in high winds, or even speaking directly into the microphone can damage them.

36 Check the manufacturers' websites for more information.

<table>
<tr><th>Brand</th><th>Model</th><th>Frequency response</th><th>Sensitivity</th><th>Directivity</th><th>Output impedance</th></tr>
<tr><td>Audio Technica</td><td>AT4081</td><td>30 Hz to 18 kHz</td><td>–42 dBV</td><td>Bidirectional</td><td>100 Ω</td></tr>
<tr><td colspan="3">A microphone with a warm and natural sound. Consists of a double ribbon structure. It withstands high sound pressure levels.</td><td colspan="3">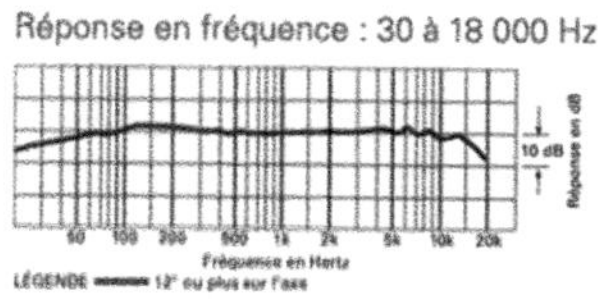
</td></tr>
<tr><td>Avantone</td><td>CR14</td><td>30 Hz to 15 kHz</td><td>–52 dBV</td><td>Bidirectional</td><td>600 Ω</td></tr>
<tr><td colspan="3">A very natural and detailed sound, very well suited for duets and vocal groups. Interesting for capturing the sound of a piano, guitar amp pickup, acoustic guitar, mandolin and banjo. Effective on a drum overhead[37].</td><td colspan="3">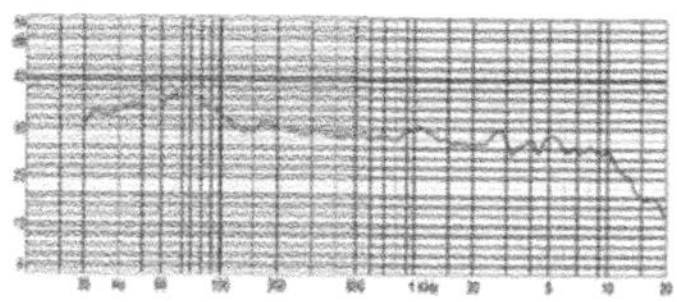</td></tr>
<tr><td>Beyerdynamic</td><td>M160</td><td>40 Hz to 18 kHz</td><td>1 mV/Pa</td><td>Hypercardioid</td><td>200 Ω</td></tr>
<tr><td colspan="3">Very good for guitar amp patching. Its directionality is very appreciable because it is not very sensitive to rear sounds. Its extra gain in the 5 kHz range gives it an excellent presence.</td><td colspan="3">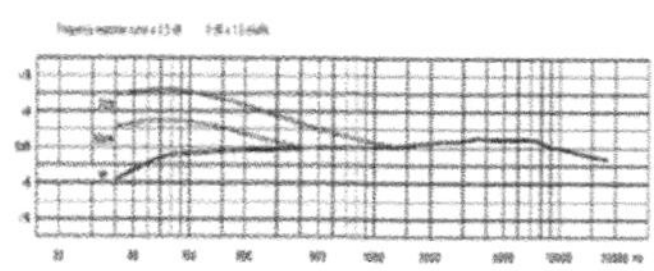</td></tr>
<tr><td>Rode</td><td>NTR</td><td>20 Hz to 20 kHz</td><td>30.5 dBV</td><td>Bidirectional</td><td>200 Ω</td></tr>
<tr><td colspan="3">A very sensitive microphone with high accuracy in the upper spectrum. Very suitable for voice recording. A robust design with shock absorber.</td><td colspan="3">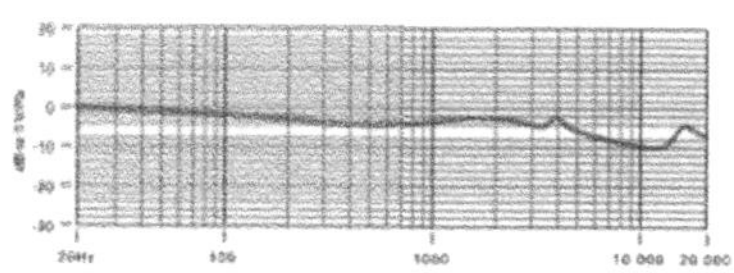</td></tr>
</table>

37 Overhead or overheads refer to the use of a pair of microphones arranged to pick up a stereophonic signal. Usually, the microphones are paired. Very much used to record a drum set. The pair is arranged in X–Y or ORTF (see Appendix 5).

<table>
<tr><th>Brand</th><th>Model</th><th>Frequency response</th><th>Sensitivity</th><th>Directivity</th><th>Output impedance</th></tr>
<tr><td>Royer</td><td>R121</td><td>30 Hz to 15 kHz</td><td>–47 dBV</td><td>Bidirectional</td><td>300 Ω</td></tr>
<tr><td colspan="3">A ribbon microphone that is a reference. It is very much used for guitar amp pickup where it provides a very detailed midrange and treble.</td><td colspan="3">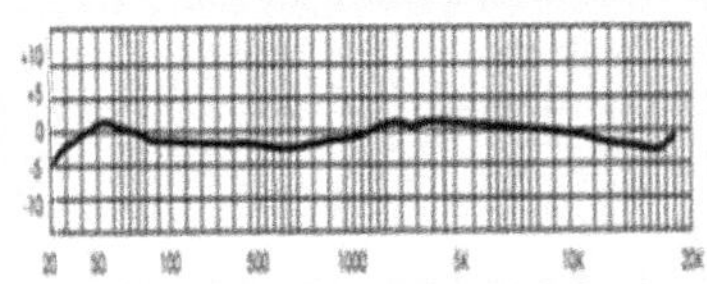</td></tr>
<tr><td>Royer</td><td>R101</td><td>30 Hz to 15 kHz</td><td>–48 dBV</td><td>Bidirectional</td><td>300 Ω</td></tr>
<tr><td colspan="3">A natural and very soft sound with very present highs. Like its big brother, the R12, it is very popular for recording electric guitars.</td><td colspan="3">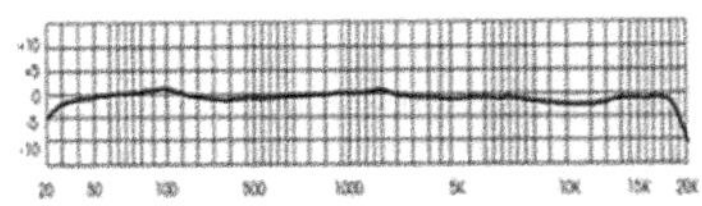</td></tr>
<tr><td>Shure</td><td>KSM353</td><td>30 Hz to 15 kHz</td><td>53.5 dBV</td><td>Bidirectional</td><td>330 Ω</td></tr>
<tr><td colspan="3">Very robust ribbon microphone withstanding high acoustic pressure. Ideal for recording fast transients like voice, as well as for acoustic instruments and concert hall recording.</td><td colspan="3">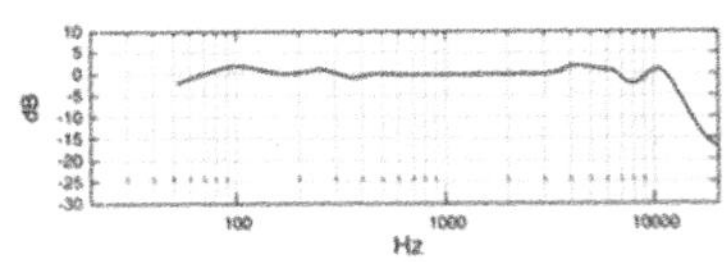</td></tr>
</table>

**Table 3.2.** *Some ribbon microphone models with their main characteristics*[38]. *For a color version of this table, see www.iste.co.uk/reveillac/recording.zip*

### 3.3.4. *Condenser microphones*

These types of microphones are known for their excellent audio quality and high sensitivity. They often capture a wider frequency range than dynamic microphones and offer a very good response to transients.

38 Check the manufacturers' websites for more information.

As mentioned above, they need to be powered, either by batteries or by phantom power.

These are most commonly used in recording studios and in environments where noise is low or can be controlled.

Their directionality can be multiple, ranging from hypercardioid to omnidirectional to bidirectional (8-shaped).

They usually have large diaphragms that are very well suited for recording voice, choir, vocals and acoustic instruments.

It should also be noted that some recent models have a USB connection and therefore a direct digital output.

They range in price from a few hundred to several thousand euros.

### 3.3.4.1. *Some condenser microphones*

<table>
<tr><th>Brand</th><th>Model</th><th>Frequency response</th><th>Sensitivity</th><th>Directivity</th><th>Output impedance</th></tr>
<tr><td>AKG</td><td>C12</td><td>30 Hz to<br>20 kHz</td><td>10 mV/Pa</td><td>Omnidirectional<br>Cardioid<br>Bidirectional</td><td>200 Ω</td></tr>
<tr><td colspan="3">Tube technology. Very suitable for vocals, brass, woodwinds, struck strings, as well as guitar or bass amp pickups. Very smooth, detailed with great presence and sensitivity.</td><td colspan="3">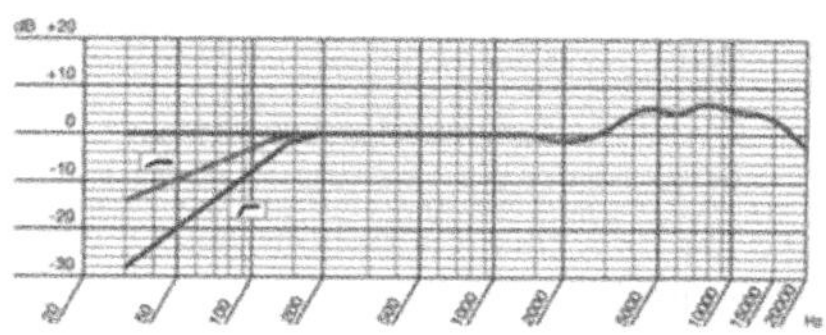</td></tr>
<tr><td>AKG</td><td>C414</td><td>20 Hz to<br>20 kHz</td><td>23 mV/Pa</td><td>9 directions</td><td>200 Ω</td></tr>
<tr><td colspan="3">Supports high sound levels with great sensitivity. A classic condenser microphone with a particular grain, especially in the midrange. Very suitable for recording acoustic instruments like strings, especially acoustic guitar.</td><td colspan="3">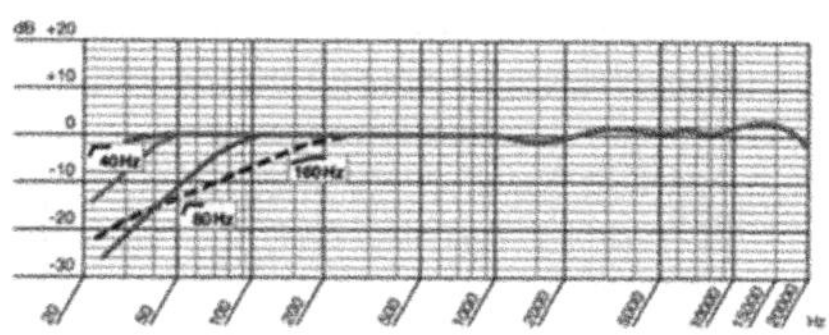</td></tr>
</table>

<table>
<tr><td>Audio Technica</td><td>AT4050</td><td>20 Hz to 18 kHz</td><td>15.8 mV/Pa</td><td>Omnidirectional<br>Cardioid<br>Bidirectional</td><td>100 Ω</td></tr>
<tr><td colspan="3">Withstands high sound pressure levels. Perfect for vocals, piano, strings, guitar amps and overheads. Suitable for stage and studio use. Equipped with an 80 Hz low-cut filter. Transformer-less technology minimizes distortion.</td><td colspan="3">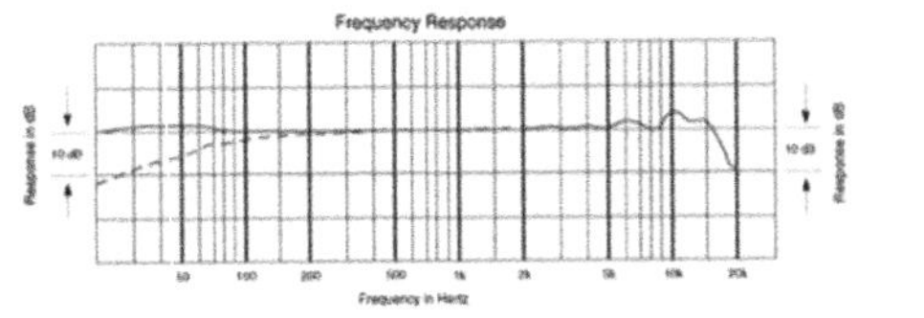</td></tr>
<tr><td>Audio Technica</td><td>AT5047</td><td>20 Hz to 20 kHz</td><td>35.5 mV/Pa</td><td>Cardioid</td><td>150 Ω</td></tr>
<tr><td colspan="3">Designed around four diaphragms to handle extreme sound pressures with high dynamics and very low noise.</td><td colspan="3">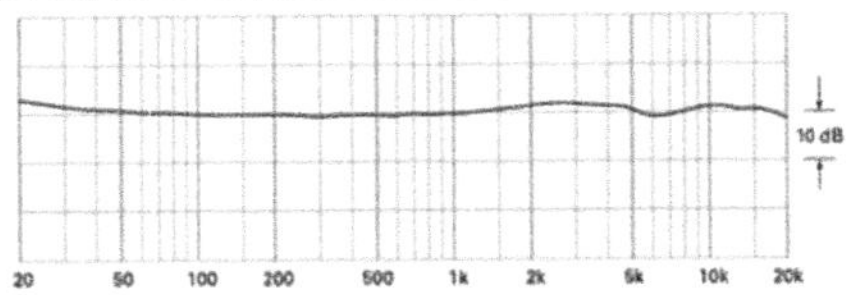</td></tr>
<tr><td>Neumann</td><td>U47</td><td>40 Hz to 16 kHz</td><td>8 mV/Pa</td><td>Cardioid</td><td>150 Ω</td></tr>
<tr><td colspan="3">A reference in condenser microphones. FET technology. A very warm sound, which is excellent for voices and singing. It supports very high acoustic pressures. Very effective for picking up guitar or bass amps and the bass drum.</td><td colspan="3">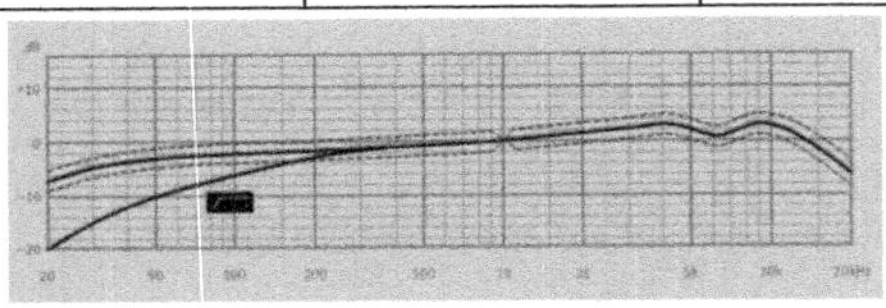</td></tr>
<tr><td>Neumann</td><td>U67</td><td>20 Hz to 20 kHz</td><td>5 mV/Pa</td><td>Cardioid<br>Bidirectional<br>Omnidirectional</td><td>200 Ω</td></tr>
<tr><td colspan="3">Like the U47, this microphone has proven itself. Tube technology. Its main quality is its universality, making it suitable for all types of recording. It is very warm and soft in the high range, with unrivaled definition. If there is one microphone that should be chosen, this is it.</td><td colspan="3">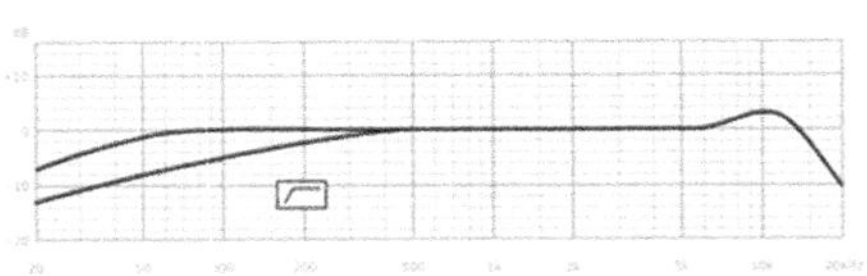</td></tr>
<tr><td>Neumann</td><td>U87</td><td>20 Hz to 20 kHz</td><td>22 mV/Pa</td><td>Cardioid<br>Bidirectional<br>Omnidirectional</td><td>200 Ω</td></tr>
<tr><td colspan="3">Like the U47 and U67, this is a must-have. A very neutral and versatile microphone. Very good on vocals, backing vocals, acoustic guitar, mandolin, banjo, violin and percussion. Great presence and very airy.</td><td colspan="3">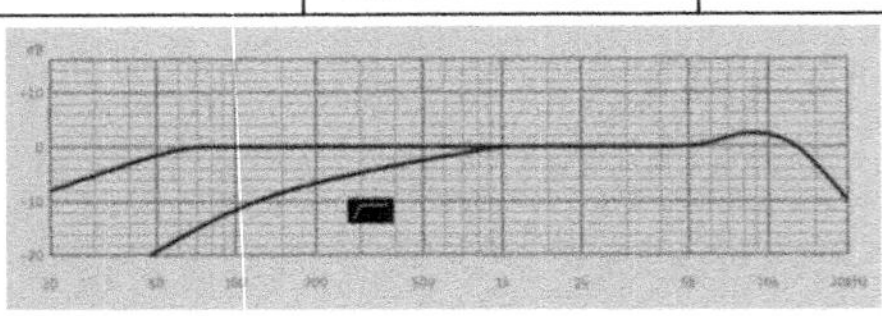</td></tr>
</table>

<table>
<tr><td>Neumann</td><td>TLM103</td><td>20 Hz to 20 kHz</td><td>23 mV/Pa</td><td>Cardioid</td><td>50 Ω</td></tr>
<tr><td colspan="3">The little brother of the U87. Low noise and high power handling. Does not contain a transformer. Great neutrality but may be sensitive to positioning. Excellent for voice and brass.</td><td colspan="3">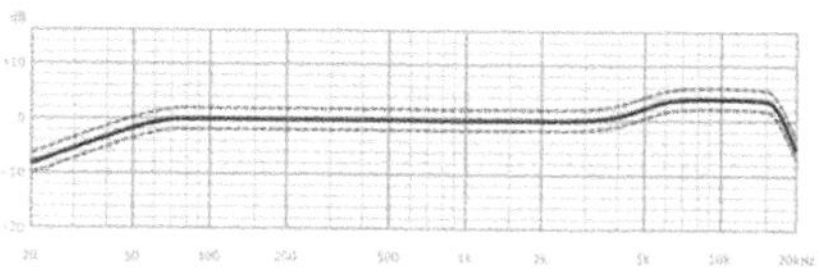</td></tr>
<tr><td>Neumann</td><td>TLM170R</td><td>20 Hz to 20 kHz</td><td>8 mV/Pa</td><td>Cardioid<br>Hypercardioid<br>Bidirectional<br>Omnidirectional</td><td>50 Ω</td></tr>
<tr><td colspan="3">A microphone with a detailed sound but perhaps a little limited on transients. Excellent low frequency response with a dense sound that is not very colorful. The various polar patterns make this microphone suitable for all situations. Very good for guitar miking, drum room and piano.</td><td colspan="3">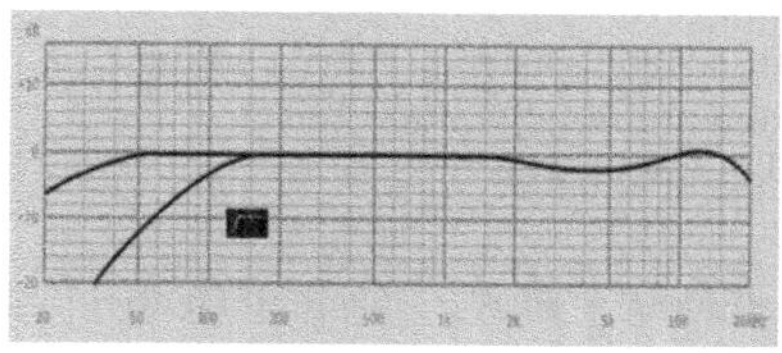</td></tr>
<tr><td>Rode</td><td>NT1</td><td>20 Hz to 20 kHz</td><td>35 mV/Pa</td><td>Cardioid</td><td>100 Ω</td></tr>
<tr><td colspan="3">A great tonal smoothness for this microphone. Very versatile and well suited for voice. Very low noise floor. A great linearity and never aggressive, even in the high end of the spectrum. An argument not to be neglected is its cost, which remains very affordable.</td><td colspan="3">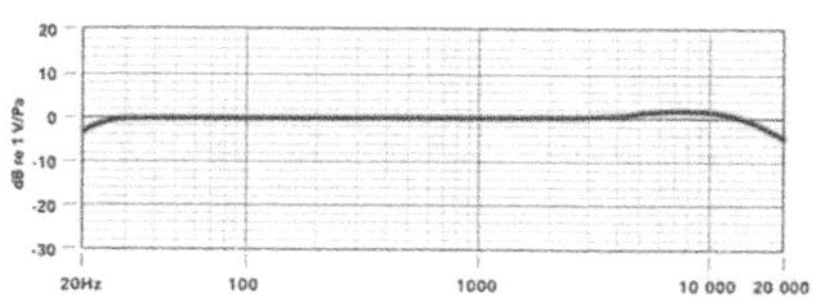
</td></tr>
<tr><td>Rode</td><td>NT2-A</td><td>20 Hz to 20 kHz</td><td>16 mV/Pa</td><td>Omnidirectional<br>Cardioid<br>Bidirectional</td><td>200 Ω</td></tr>
<tr><td colspan="3">Like the NT1, a very versatile microphone and well suited for voice with a choice of directionality. It is equipped with an adjustable low pass filter (40, 80 Hz). A good linearity gives it a clear and powerful sound.</td><td colspan="3">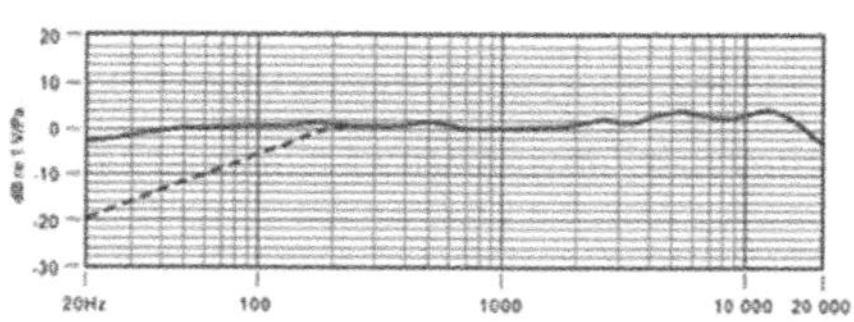
</td></tr>
</table>

| Sony | C100 | 20 Hz to 50 kHz | –31 dBV | Omnidirectional<br>Cardioid<br>Bidirectional | 90 Ω |
|---|---|---|---|---|---|
| A very accurate microphone, well suited for piano and string instruments, such as violin, cello and acoustic guitar. Very interesting and effective in its cardioid configuration. An airy side with a lot of space that is very appreciable. | | | 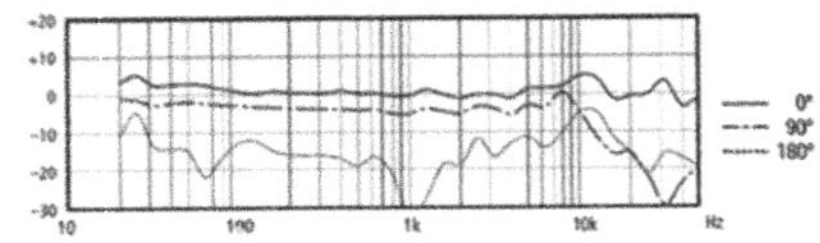 | | |
| Sony | C800G | 20 Hz to 18 kHz | –32 dBV | Omnidirectional<br>Cardioid | 100 Ω |
| Tube technology. Very popular and excellent for voice, it accepts very pronounced transients. Has a cooling system to limit distortion and background noise. | | | 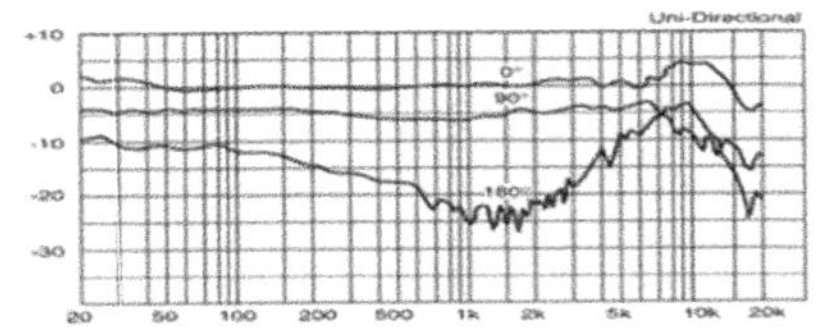 | | |
| Telefunken | ELA M251 | 20 Hz to 20 kHz | 17 mV/Pa | Omnidirectional<br>Cardioid<br>Bidirectional | 200 Ω |
| Tube technology. Perfect in overhead, especially on a battery. Very transparent and detailed. Well suited for percussion and acoustic guitar. Has an interesting and natural coloration with a low end. | | | 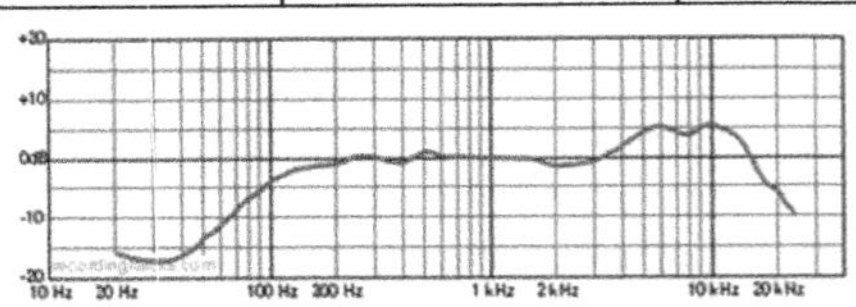 | | |

**Table 3.3.** *Some condenser microphone models[39] with their main characteristics (note: for the Neumann, Rode NT2-A, Sony, and Telefunken ELA M251 models, the frequency response curves shown are associated with the cardioid polar pattern). For a color version of this table, see www.iste.co.uk/reveillac/recording.zip*

### 3.3.5. *USB microphones*

These microphones have been designed to be used easily with any hardware or microcomputer with a USB input. They are an alternative to conventional microphones. They usually incorporate a preamplifier and an analog-to-digital converter to transform the received sound signal into a digital data stream. In terms of transducer, they work in a similar way to a standard microphone.

39 Check the manufacturers' websites for more information.

By being plug-and-play, they are very suitable for musicians and podcast creators who want to share their production directly online. They have a much higher quality than microphones built into computer equipment, and eliminate the need for a mixer, interface and traditional XLR microphones.

The converter on board the USB microphones is a very important element that conditions the quality of the digital signal obtained. We must take into account their sampling frequency and the bit depth they can work with.

The sampling frequency, in Hz, determines the number of sound samples per second – the higher it is, the more precise the signal becomes. There are four commonly used frequencies: 44.1, 48, 88.2, 96 and 192 kHz.

The depth determines the number of degrees of amplitude that each sample can support. It is measured in bits and is usually 16, 18, or 24 bits.

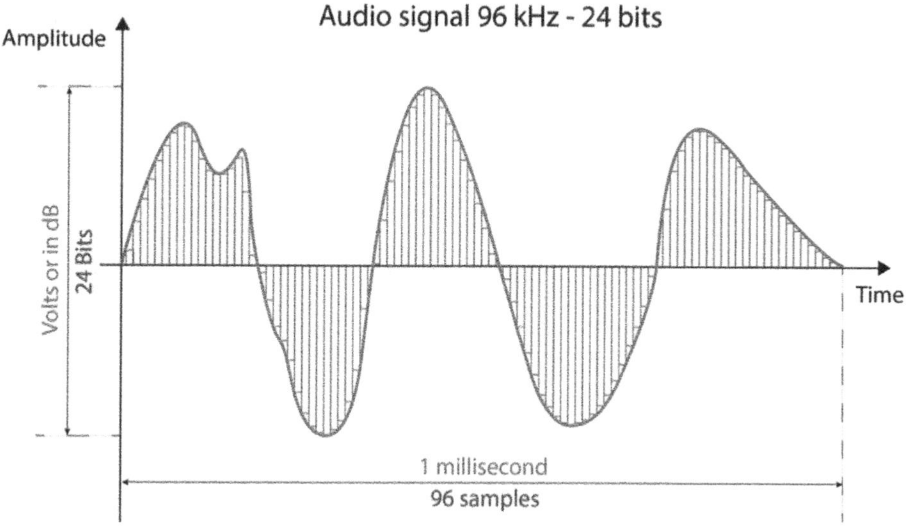


**Figure 3.45.** *A 1 ms audio signal sampled at 96 kHz. For a color version of this figure, see www.iste.co.uk/reveillac/recording.zip*

The USB microphones have a particular connectivity that varies according to the type of USB port[40]. Some have a built-in 3.5 mm headphone output and a volume control.

40 See Appendix 2.

#### 3.3.5.1. *Some USB microphones*

<table>
<tr><th>Brand</th><th>Model</th><th>Frequency response</th><th>Depth</th><th>Sample calibration</th><th>Connector</th></tr>
<tr><td>AKG</td><td>P120 USB</td><td>20 Hz to 20 kHz</td><td>24 bits</td><td>44.1–48 kHz</td><td>Micro-B to USB A</td></tr>
<tr><td colspan="6">Cardioid – Condenser – Microsoft Windows – MacOS – 20 dB Attenuator</td></tr>
<tr><td>AKG</td><td>Lyra C44 USB</td><td>20 Hz to 20 kHz</td><td>24 bits</td><td>192 kHz</td><td>USB-C</td></tr>
<tr><td colspan="6">Adjustable directionality – Condenser – Headphone output – Windows – MacOS – Android – iOS</td></tr>
<tr><td>Audio Technica</td><td>AT2020USB</td><td>20 Hz to 16 kHz</td><td>16 bits</td><td>44.1 kHz</td><td>USB-B to USB-A</td></tr>
<tr><td colspan="6">Cardioid – Condenser – Headphone output – Microsoft Windows – MacOS</td></tr>
<tr><td>Blue</td><td>Blue Yeti</td><td>20 Hz to 20 kHz</td><td>16 bits</td><td>48 kHz</td><td>Micro-B to USB-A</td></tr>
<tr><td colspan="6">Adjustable directionality – Headphone output – Microsoft Windows – MacOS</td></tr>
<tr><td>Rode</td><td>Podcaster</td><td>40 Hz to 14 kHz</td><td>18 bits</td><td>8–48 kHz</td><td>USB-B to USB-A</td></tr>
<tr><td colspan="6">Cardioid – Condenser – Headphone output – Microsoft Windows – MacOS</td></tr>
<tr><td>Rode</td><td>NT-USB</td><td>20 Hz to 20 kHz</td><td>16 bits</td><td>48 kHz</td><td>USB-B to USB-A</td></tr>
<tr><td colspan="6">Cardioid – Capacitor – Microsoft Windows – MacOS</td></tr>
<tr><td>Samson</td><td>C01U pro</td><td>20 Hz to 18 kHz</td><td>16 bits</td><td>44.1–48 kHz</td><td>Ω</td></tr>
<tr><td colspan="6">Hypercardioid – Capacitor – Headphone output – Microsoft Windows – MacOS</td></tr>
<tr><td>Shure</td><td>MV5</td><td>20 Hz to 20 kHz</td><td>24 bits</td><td>48 kHz</td><td>Micro-B to USB A or Micro-B lightning[41]</td></tr>
<tr><td colspan="6">Cardioid – Condenser – Switchable signal type (voice, vocals, instrument) – Headphone output – Microsoft Windows – MacOS – iOS</td></tr>
</table>

**Table 3.4.** *Some models of USB microphones*

41 Lightning is an 8-pin connector that has been produced by Apple since 2012.

## 3.4. Uses of microphones according to their directivity

To conclude this chapter, I will try to summarize the key points that will make it possible for you to use a particular polar pattern depending on the type of recording you want to make.

NOTE.– These are only a few tips from my modest experience, far from being gospel. Recording is above all a question of listening and feeling; everyone has their own little recipes and tricks.

### 3.4.1. *Omnidirectional microphones*

This type of microphone is theoretically designed to pick up the sound source uniformly, regardless of its position around the microphone. The coverage angle is 360°. A weak cardioid directivity may occur in the high frequencies.

Although they are less popular than cardioid microphones, they are useful in many cases:

– recording of the overall sound of a room or an environment as for a drum overhead, for example;

– recording of a moving sound source;

– recording a stereo source with a pair of microphones;

– recording of a source broadcasting a sound signal over a large width, such as a choir, an orchestra and a grand piano.

Omnidirectional microphones have a number of advantages, such as not being sensitive to proximity effect, having a wide spectrum, bringing little coloration to off-axis sources and generating little self-noise due to their small diaphragm size.

### 3.4.2. *Bidirectional microphones (figure-8)*

This type of microphone is ideal for very special situations, such as a duet of musicians or singers facing each other, but these are few and far between. On the other hand, they can be useful for recording with ribbon microphones, for recording a stereo sound signal or for isolating sounds that are out of phase with an ensemble.

The use of a ribbon microphone forces the user to consider a bidirectional directionality, which is technologically attached to their design and operation.

However, there are situations in which the bidirectional microphone is of real interest, for example, when recording stereo sound using the Blumlein system[42], in which two microphones with a figure-of-eight pattern are placed one above the other at a 90° angle. This produces a beautiful stereo image, or by switching back to mono, a signal equivalent to an omnidirectional microphone.

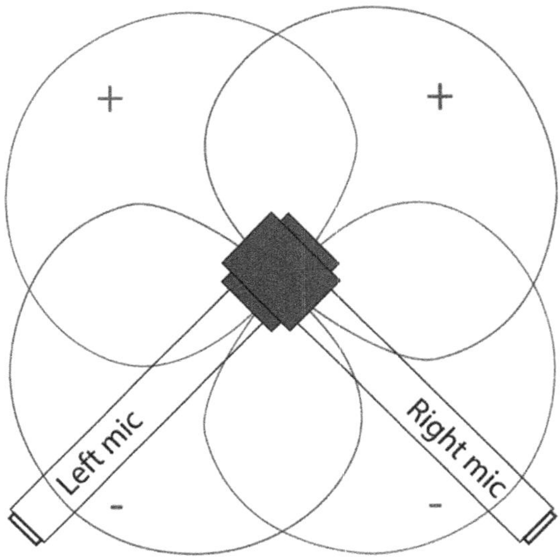


**Figure 3.46.** *Microphone configuration according to the Blumlein system. For a color version of this figure, see www.iste.co.uk/reveillac/recording.zip*

With an MS (*Mitte Seite*) configuration, which combines a cardioid or omnidirectional microphone pointed towards the center of the sound space and a microphone with a figure-8 pattern perpendicular to and as close to the former as possible. By adding and subtracting, called *matrixing*, a stereo signal is generated, the depth of which can be easily adjusted.

In certain situations, where we only want to keep the sounds placed in the axis, the figure-8 microphone has its advantages, especially since we can reduce its front or rear zone of capture by arranging acoustic absorbing panels in the axis, which will block the undesirable sound signals.

42 British sound engineer who invented this sound recording technique in 1931.

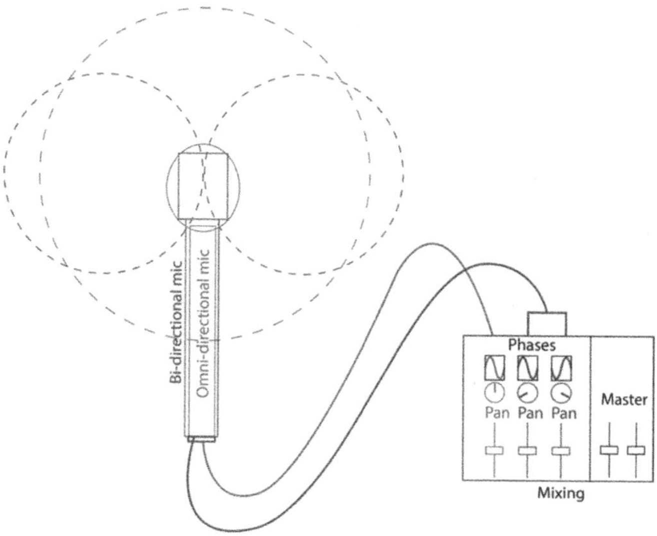


**Figure 3.47**. *Microphone configuration according to the MS system. For a color version of this figure, see www.iste.co.uk/reveillac/recording.zip*

### 3.4.3. *Cardioid microphones*

It's given that they are the most widely used, as their directionality makes it possible for them to be pointed at one sound source while ignoring all others. This is why they are the microphones of choice for voice and singing.

However, their use can present some defects that should not be neglected:

– proximity effects: when the microphone is too close to the sound source, a greater amplitude is generated in the low frequencies;

– a coloration which comes from a limitation of the sensitivity in the high frequencies when the source is not in the axis anymore.

Taking this into consideration, they are very suitable for capturing the different elements of a battery by choosing appropriate placements.

They can also limit the recording of annoying sound reflections in an environment where these cannot or are not treated.

Live and on stage, they can work wonders with proper placement while avoiding unwanted feedback effects.

The other members of the family, supercardioid, hypercardioid, subcardioid, are dedicated to more specific applications such as video and cinema, and can also find their place in a studio for very special cases.

## 3.5. Conclusion

To conclude this chapter on microphones, I would say that there are no miracle recipes for using them. The only way to master them and to ensure correct recording in any sound environment is to experiment with the equipment you have. Try to apply the little knowledge presented in this chapter while using common sense and, above all, listen carefully to the atmosphere and then to the sound after the recording.

Chapter 3 of the second volume will provide you with some additional elements for voice recording, depending on the environment and the recording conditions.

# 4

# The Acoustic Environment

In this chapter, the arrangements, devices, and all the equipment necessary for a vocal recording, whether it be for a voice-over, a singer, a group of singers or a choir, are presented.

## 4.1. Location of pickup and sound isolation

Recording voices in a studio generally requires designing the recording location. The acoustics of the room where the recording takes place have a great influence on the quality and the final sound and should not be neglected. A good microphone is useless if the conditions in the space surrounding it are not adequate.

In a closed space a sound propagates from its source in all directions; indeed, part of it circulates in a straight line up to the recording microphone and then continues until it meets a significant obstacle (a wall, a piece of furniture, an object of large proportions, a pillar, etc.) on which it rebounds until it again meets a new obstacle and so on.

Between the source and the microphone, we speak of *direct sound*; after rebound, it is the *indirect sound* or *reflected sound*. It should be noted that some of these indirect sounds are themselves recorded by the microphone.

One of the properties of direct sound is its lack of interaction with the rest of the space, meaning it is not degraded and keeps all its characteristics.

It is quite different with indirect sound, which changes after each reflection to varying degrees depending on the distance it has traveled (size of the room), the elements that it has encountered, whether they are reflective, absorbent or neutral.

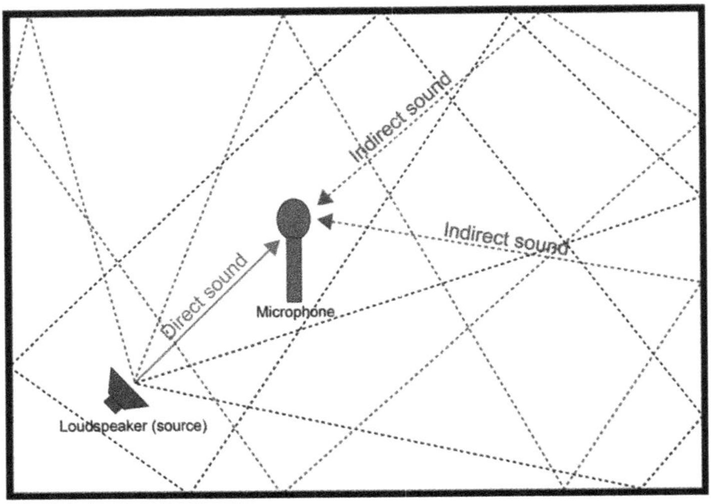


**Figure 4.1.** *Direct sound and indirect or reflected sound. For a color version of this figure, see www.iste.co.uk/reveillac/recording.zip*

In relation to what has just been said, we understand that the location, the place or the room in which the sound recording will take place is of capital importance and that it may be necessary to arrange or rework the space.

In the following sections, sound insulation is not discussed, a vast subject that would require several hundred pages to be covered correctly. In the reference list at the end of this book you will find many books and Internet links devoted to this subject; the author invites you to consult them if you are interested. Only a few acoustic processing tools will be focused on, which will enable you to improve the sound rendering for voice recording, which is the foundation of this book.

## 4.2. Acoustic processing

Here is a set of steps and solutions to transform a simple room into a well-adapted sound recording space, knowing that each case is different and that adjustments will have to be made.

### 4.2.1. *State of the art*

The first question to ask is whether the room where the recording will take place will be exclusively dedicated to the studio or whether it will integrate the mixing console, the monitoring system, electronic devices, even a sofa and some furniture.

The second question to ask is about the recording space itself. Will the room be used for a single singer, a group of singers or a choir? This will determine the minimum volume required for the speakers to be comfortable.

By means of *sound insulation*, we will try to eliminate the annoying external noises as well as or disturbing noises for the neighborhood. *Acoustic treatment* will optimize the sound behavior of the room though a clever mix of various elements, materials, acoustic traps, coatings and various panels. Insulation and treatment are two very different things but must be carried out together.

If the acoustic treatment of a place or a recording booth were to absorb all the sound signals, we would use deaf rooms and the problem would be definitively solved, but unfortunately the human ear needs a share of reflected or reverberated sounds for pleasant listening and it is there that the balance must be found.

It is important to remember that the goal is not to embellish the sound but to record it as faithfully as possible.

To establish a simple diagnosis of your premises, here is how to proceed. Clap your hands as loud as possible and listen carefully to the sound that follows. Depending on the size of the premises and the layout, you will hear a sound that will vary between two extremes: rough, dry and perhaps even metallic or rather muffled and very damped as if sucked in by the environment.

You can do this test in several locations to compare and you will see that there can be many differences.

The final objective is to obtain acoustics with the greatest possible neutrality.

This term of neutrality does not mean much and that it remains rather subjective; there are much less empirical means to arrive at this result by passing by systems of measurement and evaluation in a more scientific approach (by calculation of the RT60[1], calculation of the frequency of resonance[2], etc.) but that is not the goal of this book. The author will let you consult the reference list at the end of this book.

In the rest of this chapter, some general tips and the different necessary elements that are most used in studios to obtain a satisfactory acoustic treatment are presented.

---

1 RT60 (Reverb Time 60 dB), the reverberation time of a room for the sound pressure to decrease to one thousandth of its initial value, that is, a decrease of *60 dB*.

2 The resonance frequency, *Fr*, is equal to the celerity of sound, *c*, or about *340 m/s* divided by twice the distance of the opposite surfaces *2 x L*, or *Fr = c/2 x L*.

### 4.2.2. *Bass traps*

These elements are *acoustic panels* that absorb sound with a very high proportion of low frequencies (bass trap).

The work on the bass is one of the major points making it possible to correct the most important defects of a recording room. They exist in several forms that make it possible for them to be installed on the wall or ceiling.

Here are the different types of existing technologies for bass traps:

– porous absorber;

– Helmholtz resonator;

– membranes or flexing;

– active or electronic.

#### 4.2.2.1. *Porous absorber bass traps*

The porous absorbers are made of a material with a very high density such as acoustic foam or mineral wool, fiberglass or rock wool. The operating principle is to reduce the amplitude of the sound wave by making it travel through a set of small cavities that are connected to each other.

When the sound wave travels through these cavities, the air molecules create friction on their walls, converting the kinetic energy into thermal energy, that is, heat, while decreasing their velocity, which leads to a reduction in the intermolecular pressure, which finally reduces the amplitude of the sound wave. It is necessary to add the ability for these cavities to deform, by allowing the material to be compressible, which contributes to an even more effective absorption.

The size of the cavities for each type of material is one of the essential factors that condition the range of absorbed sound frequencies.

To be effective, this type of panel requires significant thickness and a sufficiently large space.

Generally, these panels are placed in the corners of the room. This is because the sound waves are limited by the walls, ceiling and floor, and meet in the corners, which accumulate more sound pressure, which means that placing a panel at 45° in a corner produces greater attenuation while using less space.

**Figure 4.2.** *Foam bass traps for wall corners (Auralex)*

To calculate the minimum frequency of absorption of a porous material, we apply a rule of thumb:

$$f_m = \frac{c}{4ep}$$

where $c$ is the celerity of sound, about *340 m/s* and *ep* is the thickness of the material in meters.

For example, if the thickness of the material is *30 cm*, we have:

$$f_m = \frac{340}{4 \times 0.3} = 408\ Hz$$

It is possible to artificially extend the thickness of the material by adding a *plenum* space[3]. This plenum will consist of a tight box forming a cavity filled with air and closed by the absorbent material. The thickness of the trapped air is added to the thickness of the absorbent material.

Even with this artifice, it is quickly understood that, for maximum efficiency in low frequencies, very large thicknesses are required, which is not always realistic in relation to the space available.

3 In physics, a plenum is defined as a volume of air that is pressurized relative to the outside air.

**Figure 4.3.** *Absorbing panel type bass trap (t.akustic)*

On the other hand, these bass traps also bring another disadvantage; they also trap the midrange and high frequencies, which is not necessarily the desired treatment.

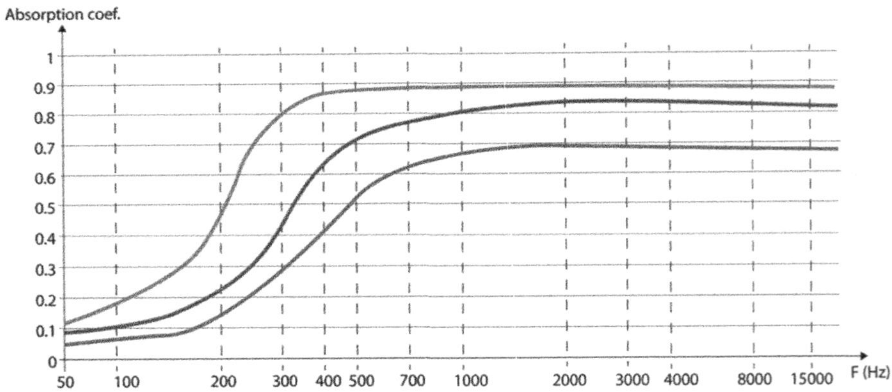


**Figure 4.4.** *Porous absorber – absorption curves (without absorber (green), absorber thickness: 20 cm (red), absorber thickness 50 cm (blue)). For a color version of this figure, see www.iste.co.uk/reveillac/recording.zip*

Porous absorbers are less expensive and simpler in design than other types, and therefore hold the largest share of the market.

### 4.2.2.2. *The Helmholtz resonator bass traps*

Contrary to porous absorbers, whose efficiency at low frequencies is low but which work well in the wide band, the Helmholtz absorbers are much more efficient at low frequencies, which are their primary aptitude, since their effects on high and medium frequencies are very negligible.

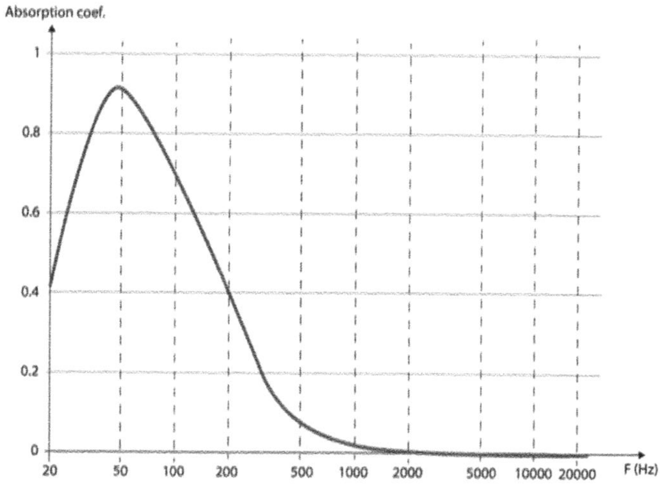


**Figure 4.5.** *Example of an attenuation curve for a Helmholtz absorber. For a color version of this figure, see www.iste.co.uk/reveillac/recording.zip*

A volume of air trapped in a cavity opened by a neck (like a bottle, for example) creates a Helmholtz resonator[4]. At a specific frequency, a resonance known as cavity resonance is created, which produces an absorption according to a very selective tuning frequency which can be calculated.

$$f_{acc} = \frac{c}{2\pi} \times \sqrt{\frac{S}{V.l}}$$

where $V$ is the volume of the cavity (in $m^3$), $S$ is the opening surface of the neck (in $m^2$), $l$ is the length of the neck (in $m$), and $c$ is the celerity of sound (340 $m/s$).

To define the principle more precisely, we can say that the air contained in the neck acts like a diaphragm, which oscillates when subjected to a sound wave. It compresses and expands the air contained in the cavity of the resonator.

4 Hermann Ludwig Ferdinand von Helmholtz, 1821–1894, physiologist and physicist who worked on the study of sound and color perception as well as thermodynamics.

But how does the system manage to massively attenuate or even suppress the sound wave that is at the resonance frequency, since the system itself produces a resonance?

Simply by a phase shift of *180°*. The sound wave emitted by the resonator is similar but in phase inversion (see Chapter 3, Figure 3.21), which produces an opposition of energy that annihilates the sound wave.

This solution is very efficient but requires making a custom-made resonator according to the frequency to be attenuated, or even several resonators to cover a determined frequency range. The cavity can contain an absorbent material to reinforce the damping of the sound waves.

These types of absorber resonators also exist prefabricated in other forms that are easier to integrate.

**Figure 4.6.** *Perforated Helmholtz panel (Acoustissimo, left) and laminar resonator (Madeinacoustic, right). For a color version of this figure, see www.iste.co.uk/reveillac/recording.zip*

Perforated panel resonators, laminar resonators and cylindrical resonators (see Figures 4.6 and 4.7).

Helmholtz absorbers are very efficient and exist for all frequency ranges, in very different shapes, aspects and dimensions from many manufacturers, at very different prices. Their strong point is their selectivity and their efficiency even in low frequencies, as opposed to porous absorbers; however, to be well adapted, they will have to be made to measure, with substantial costs if we have recourse to specialized companies.

**Figure 4.7.** *Cylindrical Helmholtz absorbers (Vicoustic and Hofa). For a color version of this figure, see www.iste.co.uk/reveillac/recording.zip*

#### 4.2.2.3. *Membrane or flexing bass traps*

Like the Helmholtz resonator bass traps, these bass traps are also very efficient and generally take up less space. Their design integrates a membrane (or diaphragm), which can be flexible (nitrile, polyurethane, vinyl, etc.) or rigid (wood, MDF, etc.).

They are watertight boxes whose front face is replaced by a material that serves as a membrane and in which a porous absorbent material, such as acoustic foam or mineral wool, has been placed.

Under the effect of the sound wave, the membrane deforms and vibrates, putting pressure on the air molecules contained in the cavity where the absorbing material slows down their speed. We thus obtain a damper that reduces the acoustic pressure by reducing the amplitude of the sound.

Some models of diaphragm bass traps incorporate a perforated panel (Helmholtz) that extends the range of absorbed frequencies.

**Figure 4.8.** *Diaphragm bass traps (GIK Acoustics). For a color version of this figure, see www.iste.co.uk/reveillac/recording.zip*

A simple mathematical approach for a diaphragm or flexural bass trap is to consider that the device is equivalent to a mass-spring system, the mass being the panel and the spring the air enclosed in the cavity. This calculation is only an approximation; indeed, the reality presents many other constraints, such as the rigidity of the surface, its deformation, its fixation on the cavity (placed, embedded, etc.) and the rigidity of the frame constituting the cavity.

To be as close as possible to the calculations that follow, the bass trap should work like a piston, that is, the membrane should be free to move back and forth by sliding in the cavity. This is physically impossible to achieve, since a joint must be made between the edges of the cavity and the membrane itself.

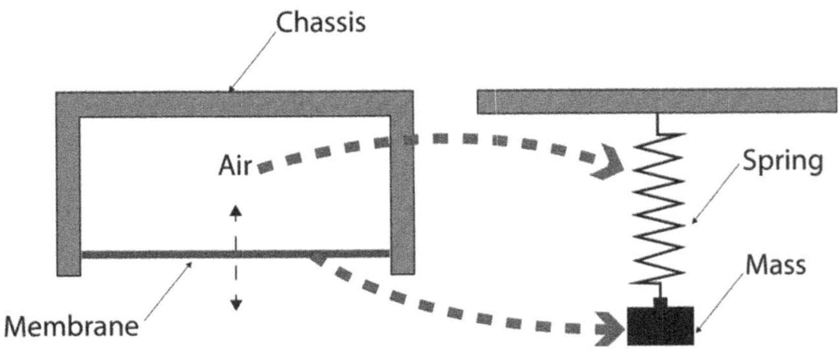


**Figure 4.9.** *Mechanical analogy for a diaphragm or flexural bass trap. For a color version of this figure, see www.iste.co.uk/reveillac/recording.zip*

Here is the calculation of the resonant frequency of a spring-mass system:

$$f_r = \frac{1}{2\pi}\sqrt{\frac{\frac{p_0 c^2}{d}}{m_s}} = \frac{c}{2\pi}\sqrt{\frac{p_0}{m_s d}}$$

where $m_s$ is the mass per unit area of the membrane, in kg/m$^2$, $p_0$ is the density of air, 1.2 kg/m$^3$ at standard temperature and pressure, $c$ is the speed of sound, 340 m/s, and $d$ is the depth of the cavity, in meters.

This formula can be simplified by integrating the constants, which leads to the formula that is generally found:

$$f_r \approx \frac{60}{\sqrt{m_s d}}$$

For example, for an MDF[5] membrane of density *800 kg/m³* of *5 mm* thickness, placed in a cavity of depth *25 cm*, we have:

$$f_r \approx \frac{60}{\sqrt{800 \times 0.005 \times 0.25}} \approx 60Hz$$

NOTE.– The mass per unit area is equal to the density in *kg/m³* by the thickness in *m*.

This type of bass trap is very effective for a reduced space, though with their construction being very delicate and the number of manufacturers being low, they are expensive without counting that most of the time they will have to be custom manufactured.

#### 4.2.2.4. *Active or electronic bass traps*

These new absorbers, which came onto the market a few years ago, are equipped with an electronic device that uses a microphone to capture the sound waves present in the room and generates a signal in a reversed phase with a loudspeaker that is supposed to destroy the selected frequencies.

They are mobile, easy to implement, small in size and are designed to simulate absorbers corresponding to 20 to 40 times their dimensions. Their major drawback is their very high cost. Only a few manufacturers design such devices.

---

5 *Medium Density Fberboard*, medium density wood fiberboard, also known as *medium*. Its main advantage is having homogeneous physical properties in three dimensions (isotropic).

**Figure 4.10.** *Two examples of active bass traps (PSI Audio – Bag End)*

### 4.2.2.5. *Manufacturers*

In Table 4.1, you will find a list of the main manufacturers of absorber panels. Some products are only available from music retailers[6].

| Manufacturer | Website |
|---|---|
| Auralex | auralex.com |
| Clearsonic | www.clearsonic.com |
| EQ Acoustics | eqacoustics.com |
| Green Acoustics | www.greenacoustics.co.za |
| Hofa | hofa-akustik.de |
| Jocavi | www.jocaviacousticpanels.com |
| Primacoustic | www.primacoustic.com |
| Ultimate Acoustics | www.ultimatesupport.com |
| t.akustic | *Reseller* |
| Vicoustic | vicoustic.com |
| Widara | *Reseller* |

**Table 4.1.** *Major manufacturers of absorbent panels (this list is not exhaustive)*

6 See the reference list at the end of the book.

### 4.2.3. *Acoustic diffusers*

While absorbers reduce the energy of sound waves in a selected frequency range, diffusers retain their energy and redistribute them based on a mathematical model in all directions to achieve a more even sound distribution in the room.

Diffusers are a complement to absorbing acoustic panels. When the defects of the room in terms of reverberation are corrected, diffusers will try to restore the sound in a more natural way and make it more alive, more musical, more intelligible and more dynamic. They are particularly effective on high frequencies.

An acoustic diffusion panel refines the sound image by avoiding certain undesirable effects; it has significant advantages:

– comb filtering removal;

– equal distribution of sound energy;

– space expansion;

– homogenization of the frequency response;

– suppression of unwanted echoes.

#### 4.2.3.1. *Suppression of comb filtering*

In a room, a wall, depending on its structure (brick, plasterboard, concrete, etc.) reflects 80 to 90% of the sound energy it receives. Any reflecting flat surface produces waves reflected at the same angle as the *incidence*. These reflections are called *specular*.

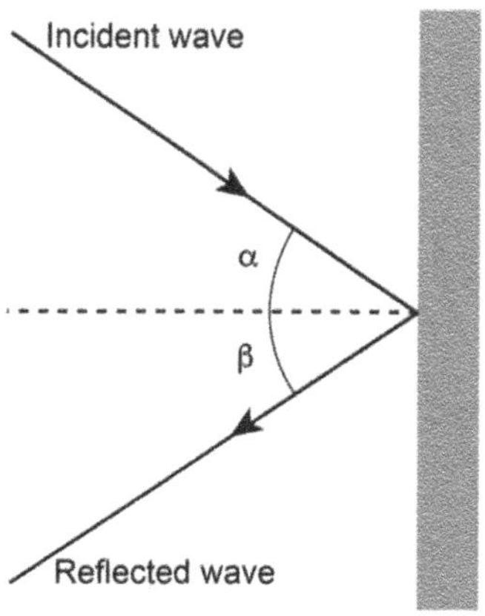


**Figure 4.11.** *A specular reflection, the angle of the incident wave is equal to that of the reflected wave ($\alpha = \beta$)*

At the point where the microphone picks up the sound, an almost identical signal will be added with a longer delay, due to the distance the sound wave will have taken to travel from the point of emission to the wall and back.

If the duration of this path is less than 30 ms (i.e. a distance of less than 10 m), a frequency distortion of the audio signal is created, called *comb filtering*. It creates peaks and troughs in the spectrum.

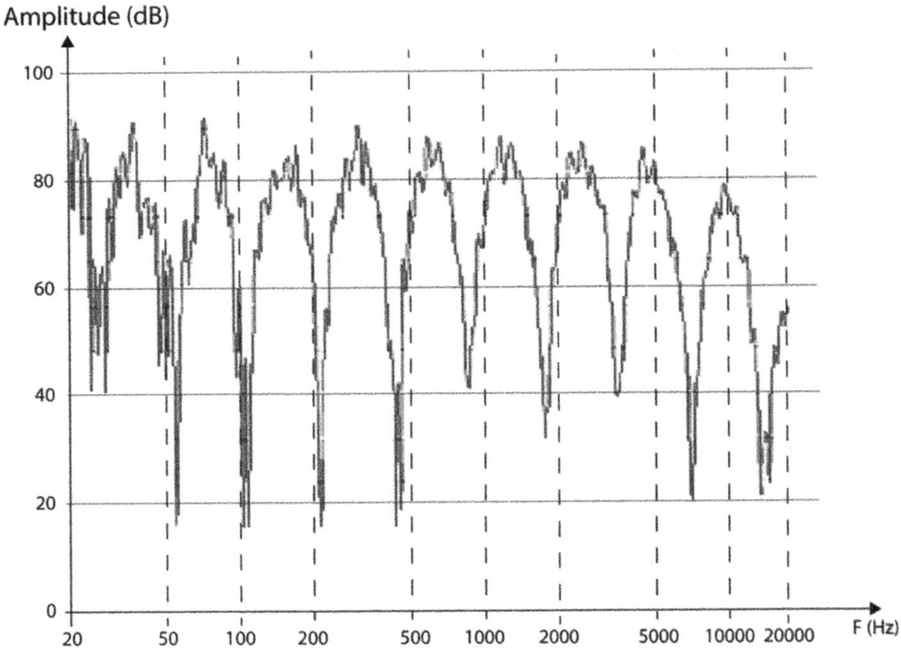


**Figure 4.12.** *Comb filtering between 20 and 20 kHz. For a color version of this figure, see www.iste.co.uk/reveillac/recording.zip*

After 30 ms, we start to perceive an echo and the frequency distortion disappears.

A diffuser will destroy specular reflections and reflect the sound energy with the same intensity in all directions.

#### 4.2.3.2. *Equal distribution of sound energy*

One of the main qualities felt in a listening place is the feeling of being part of the sound space.

Placing diffusers on the side reinforces this perception and this feeling of envelopment.

Mr. R. Schroeder[7] has worked a lot on this subject and has demonstrated the importance of the immersive aspect during music listening. To promote this immersion, he invented a system of diffusion of the sound waves, called a *Schroeder diffuser*.

**Figure 4.13.** *Two Schroeder diffusers. For a color version of this figure, see www.iste.co.uk/reveillac/recording.zip*

### 4.2.3.3. *Enlargement of the space*

Quadratic residue diffusers, also called QRDs or PRDs (Primitive Root Diffusers), are built with elements of different heights that chaotically reflect the sound waves they receive and contribute to the homogenization of the energy dispersion. They also influence the diffusion by generating very small delays and time shifts. The combination of these two factors smoothens the reverberation decay curve of a room for all frequencies taken into account by the diffuser.

7 Manfred Robert Schroeder, 1926–2009, German physicist known for his work on acoustics, especially on resonant cavities and computer graphics.

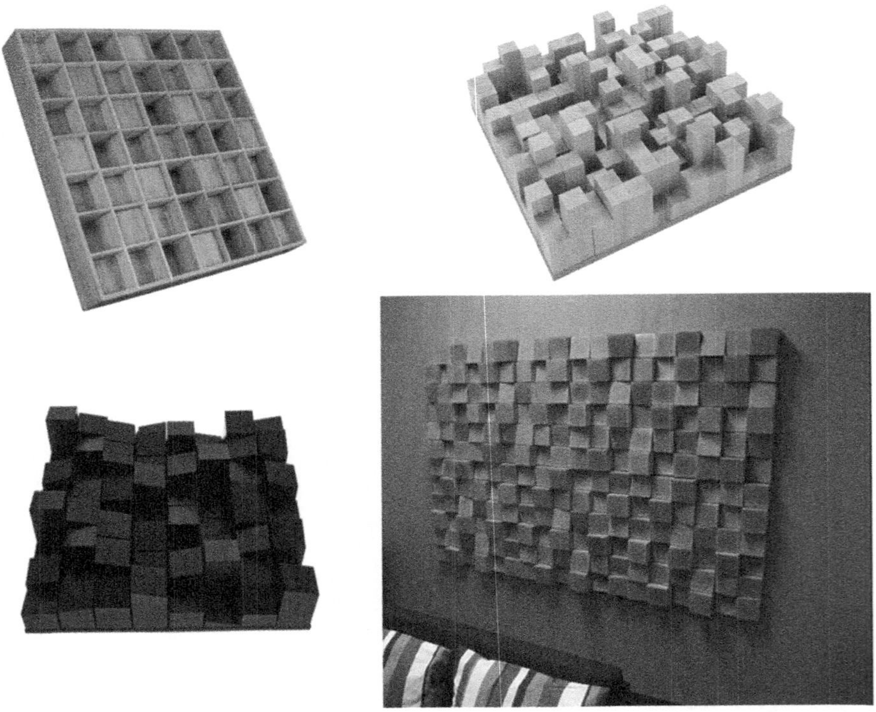

**Figure 4.14.** *Quadratic diffuser, also called QRD diffuser (Auralex, t.akustic and Vicoustic). For a color version of this figure, see www.iste.co.uk/reveillac/recording.zip*

This parameter is one of the keys that characterize large spaces, so an excellent diffusion of sound waves gives a room an acoustic dimension much larger than it really is.

These types of diffusers are used extensively in small studios with low ceilings.

### 4.2.3.4. *Homogenization of the frequency response*

In a room, an audio signal is not perceived equally at all points of the space; if you get closer to the walls or you stand in the corners, the bass often becomes more present, not to mention other phenomena that may occur.

Having a homogeneous response in the entire space quickly becomes essential, making it easier, for example, to find the ideal position to place a microphone.

The installation of diffuser panels makes it possible to eliminate parasitic reflections, to keep the naturalness of the sound spectrum without bringing coloration to it.

Another strong point is the neutrality and the homogeneity of what is heard if several people are placed in the studio at different places, they will all hear the same thing with an identical sound rendering.

In addition, the standing waves coming from the various reflected sound waves meeting often bring phenomena of amplification or destruction within the room; the diffuser is also there to attenuate them.

#### 4.2.3.5. *Suppression of unwanted echoes*

In untreated rooms, we often find *flutter echo*. This is a phenomenon that sends sound waves back and forth to infinity between parallel surfaces like two mirrors facing each other.

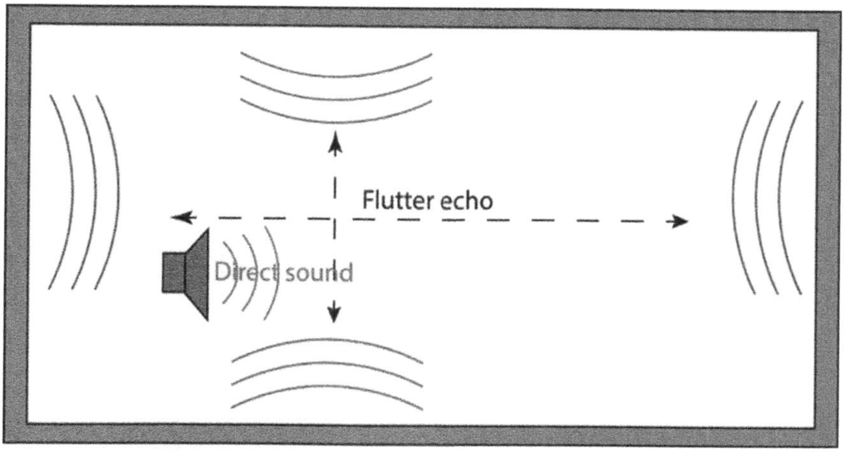


**Figure 4.15.** *The flutter echo principle. For a color version of this figure, see www.iste.co.uk/reveillac/recording.zip*

Diffusers can easily make the parallelism of the walls disappear and avoid the creation of a parasitic note that is positioned in the frequency spectrum. This sound artifact is created by the repetition of a very brief noise that gives the illusion of a continuous sound.

#### 4.2.3.6. *To go a bit further*

To support what was said earlier about diffusers, there are three main types of diffusers:

– Schroeder's diffusers;

– quadratic diffusers;

– poly-cylindrical diffusers.

Schroeder's diffusers are made of slats, usually of wood, of different widths arranged vertically, which are intended to disperse sound waves laterally.

Like Schroeder's diffusers, quadratic diffusers generally present cubes of different heights that disperse the sound in a chaotic way. They are mostly made of wood. The more differences there are between the heights of the cubes, the more the diffuser works at the low end of the spectrum.

The poly-cylindrical diffuser also spreads the sound in all directions, thanks to its rounded shape.

**Figure 4.16.** *A poly-cylindrical diffuser (Acoustic Geometry and GIK Acoustics). For a color version of this figure, see www.iste.co.uk/reveillac/recording.zip*

There are variants combining several types with particular shapes. In terms of materials, while many diffusers are made of wood (solid, plywood, MDF, etc.), there are some in less noble materials such as plaster, melamine, aluminum and acrylic glass.

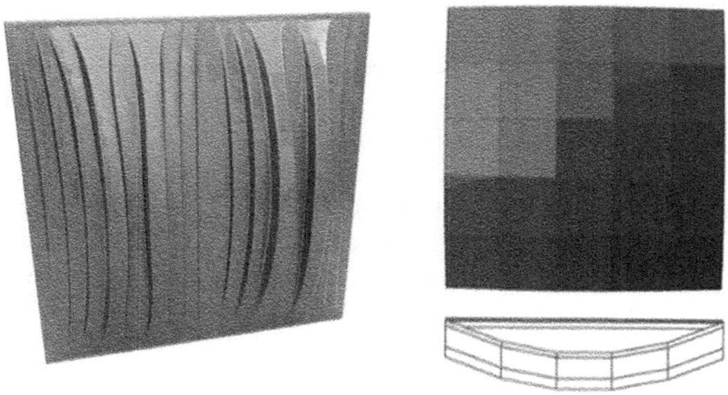

**Figure 4.17.** *Diffusers with original designs (Jocavi and Jaya). For a color version of this figure, see www.iste.co.uk/reveillac/recording.zip*

As far as their installation is concerned, they are usually placed on the wall or ceiling. In studios, they are often placed behind the instrumentalists or singers, or even laterally, although an important point must be respected: they must be at a sufficient distance, at least 1.5 meters, or even more, depending on their dimensions, so that they reach their full effectiveness.

Table 4.2 lists the major manufacturers of diffusers. Many bass trap manufacturers also design diffusers. Most products are available from major music retailers[8].

| Manufacturer | Website |
|---|---|
| ATS Acoustic | www.atsacoustics.com |
| Auralex | auralex.com |
| BXI | *Reseller* |
| Hofa | hofa-akustik.de |
| Jacovi | www.jocaviacousticpanels.com |
| Mega Acoustic | www.mega-acoustic.pl |
| MSR Acoustic | msr-inc.com/ |
| Primacoustic | www.primacoustic.com |
| Ultimate Acoustics | www.ultimatesupport.com |
| t.akustic | *Reseller* |
| Vicoustic | vicoustic.com |

**Table 4.2.** *Major manufacturers of diffusers (this list is not exhaustive)*

8 See the reference list at the end of the book.

## 4.3. Acoustic booths

For podcast creators, interpreters, voice-over commentators, whether in a home studio or a classical recording studio, it is in their interest to have perfect acoustics, which is not always possible due to the configuration of sound recording rooms.

It is possible to turn to an acoustic booth designed with absorbing panels and reflecting panels, which will control the crosstalk[9] and reflections, and ensure, at the same time, acoustic insulation. Inaudible ventilation systems are often provided to ensure air circulation inside the booth.

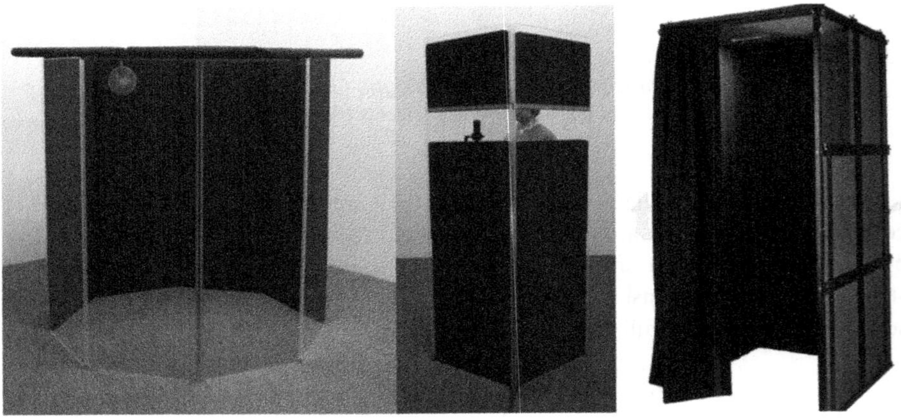

**Figure 4.18.** *Examples of removable cabins (Clearsonic and t.akustic)*

Several manufacturers have cabins of comfortable dimensions, easy to assemble and quickly dismantled. There are also smaller booths reserved for singing.

Solutions with removable panels mounted on a tripod are easy to install and can provide an interesting acoustic treatment under certain conditions.

For long-lasting installations, it is possible to buy a prefabricated cabin, with a door, ventilation, floor shock absorbers, and many other possible options, or to go through a builder who will come and create a custom cabin for you. The costs will be significant; however, it is a solution that can be interesting. The investment in a prefabricated or dismountable cabin is often as expensive.

9 Interference of a first signal with a second.

**Figure 4.19.** *A cabin with removable panels on a tripod (Power Studio)*

**Figure 4.20.** *Premanufactured acoustic booths (Vicoustic and Keoda). For a color version of this figure, see www.iste.co.uk/reveillac/recording.zip*

Table 4.3 shows you some of the manufacturers of acoustic booths. Some of them do not have a website; therefore, you have to look for their products at dealers[10].

10 See list in the reference list at the end of the book.

| Manufacturer | Website |
|---|---|
| Clearsonic | www.clearsonic.com |
| Demvox | www.demvox.com |
| Keoda | www.keoda.com |
| Oco Silence | www.oco-silence.com |
| Power Studio | *Reseller* |
| SQC Acoustics | *Reseller* |
| t.akustic | *Reseller* |
| Vicoustic | vicoustic.com |

**Table 4.3.** *Major manufacturers of acoustic booths (this list is not exhaustive)*

## 4.4. Accessories

After having reviewed the various elements necessary for correct acoustic treatment in terms of furniture, some additional accessories which are often useful or even essential are presented.

### 4.4.1. *Acoustic shields*

These devices are placed behind the microphone. They can be mounted on a standard microphone stand—provided that it is sturdy, as these devices are usually quite heavy—or on a dedicated tripod. Tabletop versions are also available.

**Figure 4.21.** *Three examples: a tripod (Neewer), microphone stand (Tonor) and tabletop (Devine RF5) isolation shields*

The main function of these shields is to help produce studio quality sound in an uncontrolled environment, as one might do in a recording studio.

Typically, these shields contain an acoustic foam similar to that used in porous absorber bass traps. The purpose of the foam is to absorb ambient noise that might reach the microphone.

This foam is placed inside a rigid shell, in the middle of which, a few centimeters away, the suspension system and the microphone for the sound recording are placed.

The shell has a more or less rounded shape or a multi-panel articulated system that can be adjusted to limit the echoes generated by the voice of the commentator or singer.

**Figure 4.22.** *A sound barrier and a sound recording session (source: www.neewer.com)*

The choice of an effective isolation shield is not simple, it is necessary to take into account several criteria, which are related to the type of use (narration, singing, commentary, etc.), the place (unprepared room, studio, office, etc.), the sound environment, the position of the user (standing, sitting), the size of the microphone and its suspension and whether it needs to be portable or will be fixed.

The main characteristics of an isolation shield are as follows:

– the dimensions that will determine the size and, generally, the type of microphone that can be installed, mainly the height and depth;

– the weight, an important criterion that will determine the choice of the final support (tripod, stand, etc.) if it is not integrated;

– the type of fixation, on foot, integrated, on table, universal;

– the quality of the porous absorber(s), which will determine the isolation of the microphone from the surrounding sound environment;

– adjustments, variable spacing and orientation, joints;

– the ease of disassembly and storage for mobile applications, which will determine its portability;

– the material of the shell, which affects the general weight, the robustness, and the absence of sound reflectivity.

In Table 4.4, you will find a list of the main manufacturers of isolation shields and their website addresses. Some of them do not have a website; therefore, you have to look for their products at resellers[11].

| Manufacturer's name | Website |
| --- | --- |
| Aston | www.astonmics.com |
| Auralex | auralex.com |
| Eagletone | www.eagletone.com/ |
| Marantz | www.marantzpro.com/products/browse/category/studio |
| Moukey | *Reseller* |
| Neewer | neewer.com/collections/microphones-accessories |
| OurLeeme | *Reseller* |
| Primacoustic | www.primacoustic.com |
| Pronomic | *Reseller* |
| RD Acoustic | www.rdacoustic.cz |
| Rode | www.rode.com/ |
| sE Electronics | www.seelectronics.com |
| t.akustic | *Reseller* |
| Tonor | www.tonormic.com |
| Vicoustic | vicoustic.com |

**Table 4.4.** *Major manufacturers of isolation shields (this list is not exhaustive)*

11 See list in the reference list at the end of the book.

### 4.4.2. *Pop filters*

When you want to record a voice, a pop filter is a must. It is designed around a fine mesh net or grid and is placed between the microphone and the mouth of the commentator or performer.

There are many filters on the market, and it is important to really know its usefulness to choose correctly.

**Figure 4.23.** *A universal pop filter*

Its main role is to block the plosives, also called *occlusive* consonants, that is, the mouth noises generated by the pronunciation of the letters "p", "b", "t", as well as to reduce possible wind noises and to block the projections of saliva that could damage the capsule of the microphone.

This is a widely used tool that attaches to the microphone suspension, stand or boom.

Not all filters are equivalent and they have their own characteristics, notably in terms of grid, shape, arms (length, joints, gooseneck, etc.), fixing system and robustness.

In Table 4.5, you will find a list of the major pop filter manufacturers and their website addresses. Many microphone manufacturers make their own pop filters.

Some of them do not have a website, so you have to look for their products at resellers[12].

12 See list in the reference list at the end of the book.

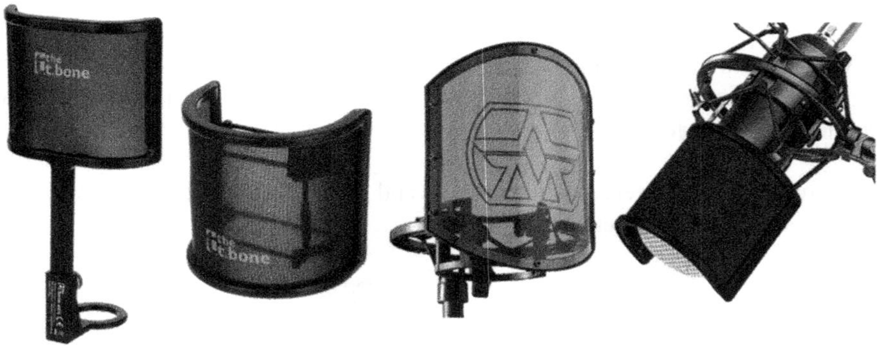


**Figure 4.24.** *Pop filters of various shapes (t-Bone, Aston and Tuloka)*

| Brand | Website |
|---|---|
| AKG | www.akg.com |
| Aston | www.astonmics.com |
| InnoGear | www.innogear.com |
| JZK | *Reseller* |
| König and Meyer | www.k-m.de |
| Neewer | neewer.com/collections/microphones-accessories |
| Neumann | www.neumann.com |
| Pauly | www.paulyton.de |
| Samson | www.samsontech.com/ |
| sE Electronics | www.seelectronics.com |
| Sennheiser | www.sennheiser.com |
| Shure | www.shure.com |
| Sywon | *Reseller* |
| Rycote | rycote.com |
| Tuloka | *Reseller* |

**Table 4.5.** *Major manufacturers of pop filters (this list is not exhaustive)*

### 4.4.3. *Headphones*

Headphones are essential equipment in a studio, both for listening to a mix or a *mastering*, and also during sound recording, for example, when the singer comes to superimpose their voice (*backing track*) on the musical arrangement, which is commonly done.

There are several hundred models on the market with a very wide price range, from a few dozen euros to several hundred.

In fact, there are two main types: closed headphones, dedicated to recording and sound recording, and open headphones, used for mixing and mastering. A less common hybrid type called semi-open also exists.

Closed-back headphones offer excellent sound isolation, cutting the listener off from the outside sound environment and also preventing unfortunate feedback into the microphone during recording (*leaking*). The musician or singer can listen to the mix and be sure that it will not interfere with the recording. This type of headphone is, therefore, very useful for monitoring.

Open headphones give a better stereo image, and you can better position the instruments in the space, but generally weaken the low end of the spectrum. In addition, they are very permeable to the outside sound environment and easily leak sound.

To recognize between the types, it is quite simple. Just check for the presence of a grid on the outside of the earpieces, which exists, most of the time, on an open headphone.

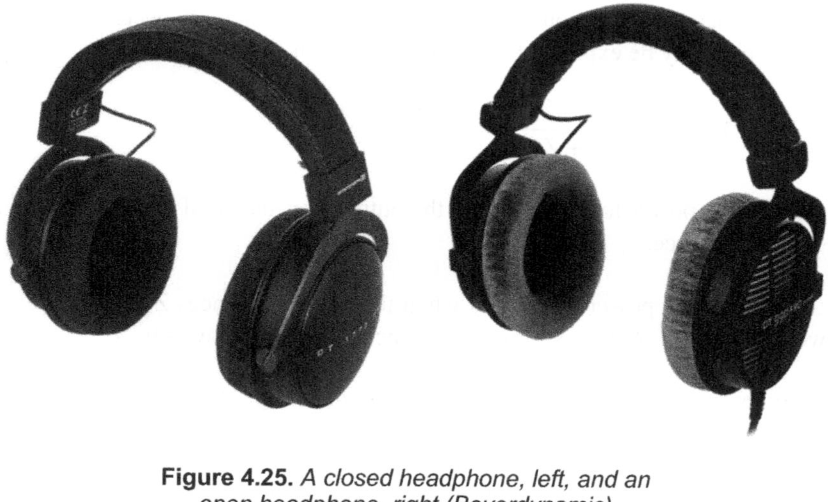

**Figure 4.25.** *A closed headphone, left, and an open headphone, right (Beyerdynamic)*

In addition, depending on the shape of the earpieces, the headphones can be *supra-aural* or *circumaural*. In the former case, the earpieces rest on the cartilage of the ears and the ear cushions serve as protection. The pressure exerted by the headband determines the comfort over time and the acoustic isolation from the environment. In the second model, circumaural, the headphones cover the ears and their cushions rest directly on the skull. These headphones are generally more comfortable if the pressure of the ring is correct. Their acoustic insulation is very good, but they remain more imposing in terms of space than their brother, the supra-aural. Attention to wearers of glasses, who can be bothered by the cushions and the pressure exerted, is to be taken into account.

We cannot talk about headphones without talking about impedance, which is one of the main parameters of headphones. If the impedance is low, less than 50 ohms, it will be easily usable on a device with low amplification. If the impedance is high, above 200 ohms, a strong signal is needed to make them work properly.

Besides the impedance, another characteristic can also appear: the sensitivity in dB SPL/mW[13].

Impedance defines the opposition that an electrical circuit offers to the passage of an alternating current. The greater the impedance, the greater the opposition.

The impedance of headphones usually varies between 8 and 600 ohms, and is determined by the nature of the speaker coils located in each of the headphones.

When a headphone is connected to a sound source, there is necessarily a power loss that can easily be expressed with the following formula:

$$P = 10 \times \log(\frac{Z_S \times Z_C}{(Z_s + Z_c)^2}$$

where $P$ is the power loss in $dB$, $Z_s$ is the output impedance of the source, and $Z_c$ is the load impedance.

The maximum power is reached when the two impedances $Z_s$ and $Z_c$ are equal. However, a loss of *-6 dB* is still present, which divides the power by *4*.

Another phenomenon to consider is the damping of the speakers. When a headphone receives a sound signal, the loudspeaker diaphragms vibrate. If the sound stops abruptly, the vibration continues, which affects the accuracy of the audio response. An electronic or mechanical device is present to dampen the displacement

13 See Appendix 1.

and compensate for this effect. In order for this device to be effective, it is necessary to ensure that the output impedance is lower than the impedance of the headphones.

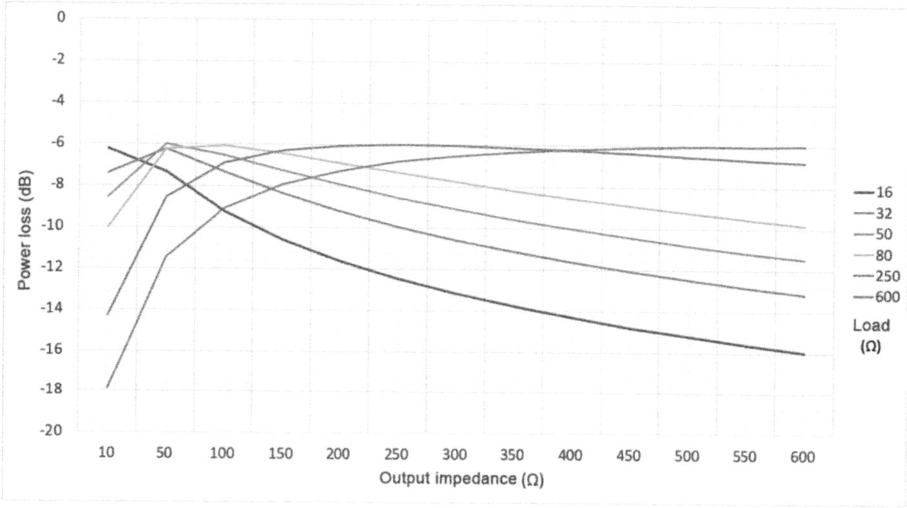


**Figure 4.26.** *Power loss (attenuation) curve according to amplifier output impedance and headphone load impedance. For a color version of this figure, see www.iste.co.uk/reveillac/recording.zip*

It should be noted that, in general, it is advisable to choose headphones with an impedance eight times higher than the output impedance of the amplifier, knowing that, in fact, the impedance of the headphones varies according to the frequency of the audio signal they receive.

Previously, the author spoke about the sensitivity of headphones, which is another important parameter to consider. This represents the sound volume emitted by the headphones for a defined power level.

The impedance and sensitivity parameters are complementary; if two headphones have the same impedance, but one of them is more sensitive, then, for the same power emitted by the amplification, the more sensitive one will emit a greater sound volume.

The unit that measures the sensitivity is the decibel per milliwatt. It is common for a headphone to have a sensitivity between *80* and *120 dB SPL/mW*.

The choice of sensitivity affects the operation of the amplifier. The greater the sensitivity, the less the output power must be important and vice versa. A headphone should be able to deliver about *110 dB peak*. At this level, the human ear can suffer damage, which is why it should only be on peak signals of very short duration, a few seconds at most. The comfortable listening level is instead in the range of *65* to *90 dB SPL*.

To calculate the power required according to the sensitivity, you can apply the following formula:

$$P = 10^{\frac{N_C - s}{10}}$$

where $P$ is the power in $mW$, $N_c$ is the level in $dB\ SPL$ sought, and $s$ is the sensitivity in $dB/mW$.

Figure 4.27 shows the results of this calculation for *100*, *105*, *110* and *115 dB SPL*, knowing that *110 dB SPL* can be considered as the reference, in relation to what the author said above.

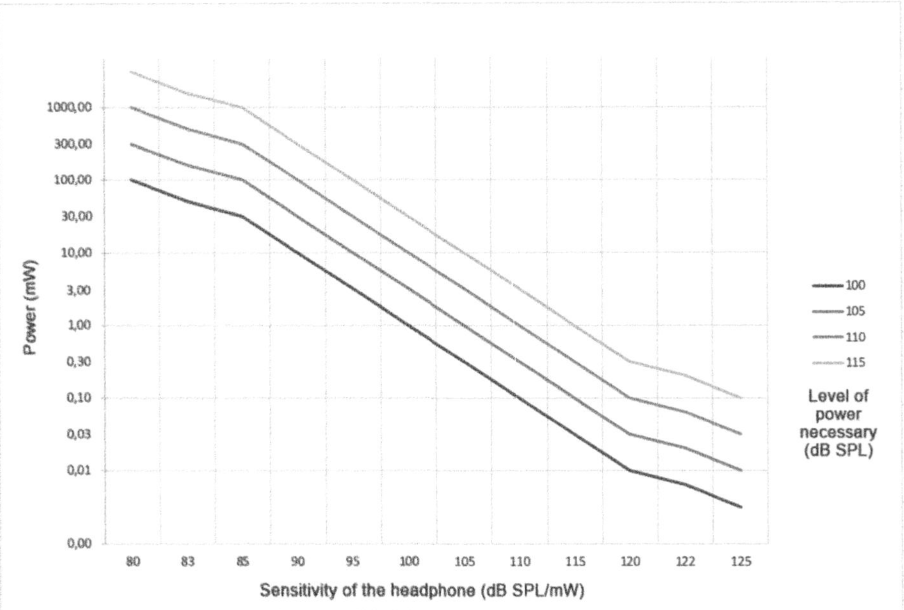


**Figure 4.27.** *Graph for determining the power required for a headphone based on its sensitivity. For a color version of this figure, see www.iste.co.uk/reveillac/recording.zip*

For example, for a sensitivity of *100 dB SPL/mW*, a power of *10 mW is needed* (referring to *110 dB* peak).

To conclude, it is necessary to pay particular attention to the impedance and sensitivity of a pair of headphones in order to be sure that the source, the output amplifier, is powerful enough to allow for listening in the best conditions. Also consider the bandwidth, which can cover a wide range.

Table 4.6 lists some major headphone manufacturers. They usually have both open and closed models. For each of them, you will find a few popular models, although this list is far from exhaustive.

| Manufacturer | Models – features | Impedance | Bandwidth |
|---|---|---|---|
| AKG | K182 – Closed, Circum<br>K371 – Closed, Circum<br>K612 Pro – Open, Circum<br>K702 – Open, Circum<br>K712 Pro – Open, Circum | 32 Ω<br>32 Ω<br>120 Ω<br>62 Ω<br>62 Ω | 10–28,000 Hz<br>5–40,000 Hz<br>12–39,500 Hz<br>10–39,800 Hz<br>10–39,800 Hz |
| Audio-Technica | ATH-M50X – Closed, Circum<br>ATH-R70X – Closed, Supra | 38 Ω<br>38 Ω | 15–28,000 Hz<br>5–40,000 Hz |
| Beyerdynamic | DT770 Pro – Closed, Circum<br>DT880 Pro – Semi-open, Circum<br>DT990 Pro – Open, Circum | 250 or 80 Ω<br>250 Ω<br>250 Ω | 5–35,000 Hz<br>5–35,000 Hz<br>5–30,000 Hz |
| Focal | Listem Pro – Closed, Circum | 32 Ω | 5–22,000 Hz |
| Sennheiser | HD26 Pro – Closed, Circum<br>HD200 Pro – Closed, Circum<br>HD300 Pro – Closed, Circum | 100 Ω<br>32 Ω<br>64 Ω | 20–18,000 Hz<br>20–20,000 Hz<br>6–25,000 Hz |
| Shure | SHR440 – Closed, Circum<br>SRH1540 – Closed, Circum | 44 Ω<br>46 Ω | 10–22,000 Hz<br>5–25,000 Hz |
| Sony | MDR-7506 – Closed, Circum<br>MDR-7510 – Closed, Circum | 63 Ω<br>24 Ω | 10–20,000 Hz<br>5–40,000 Hz |

**Figure 4.6.** *Some headphone models*

When choosing a pair of headphones, think about whether the cord and the earpads are detachable or interchangeable. These are elements that are often put to the test and it is very unpleasant to throw away a pair of headphones simply because the cord is cut or the earpads have deteriorated. Also pay attention to the cord's length and the connector that ends it, mainly on the source side (amplifier), whether it is a 3.5 mm or 6.35 mm (1/4'') jack or mixed, with an adapter provided.

### 4.4.4. *Microphone suspensions*

The suspension is a shock-absorbing device for holding the microphone, which is often indispensable for vocal recording in the studio.

It avoids the parasitic noises linked to vibrations which could occur at the level of the foot or the arm of support by mechanically damping them, generally with flexible elements, elastic cords, rubber, flexible plastic lyres, etc.

**Figure 4.28.** *Different types of suspensions: spider, double-lyre, universal, buoy (Neumann, Rode, Prodipe and Shure)*

Many high-end microphones are usually supplied with suspensions, but this is not a general rule. There are many different models (spider, double-binder, universal, buoy, etc.) depending on the nature and size of the microphone.

### 4.4.5. *Feet, poles, and arms*

Microphone suspensions or clamps can be screwed onto floor stands, wall mounts, or table or desk mounts. On the other hand, it can be useful to have a lateral offset to bring the microphone to the right place without being disturbed by the environment: instrument, amp, furniture, etc.

There are many solutions: straight microphone stands, telescopic or non-telescopic boom stands, *overhead*[14] stands, table stands, goosenecks and articulated arms.

Straight microphone stands are different depending on their manufacture. They can be entirely made of steel, from steel and plastic, with a folding tripod, with a fixed tripod, with a cast iron base, and made of rods of varying diameters. Two characteristics are to be taken into account: the weight and the mini–maxi height. A heavy stand with large diameter tubes is often a guarantee of quality.

**Figure 4.29.** *Examples of stands*

Many stands include a tilting boom that makes it possible for the microphone to be moved out of the way. The length of the boom can vary and it can be telescopic. An important point to check is the clamping device at the joint, which must be able to support the weight of the microphone with its clamp, support, or suspension without problems. In the case of great length and/or height (overhead stand), the boom can be equipped with a counterweight to balance the device (metal mass or sandbag for great lengths). A pole can be purchased independently to be added to a straight leg.

14 An *overhead* recording positions the microphones high up in relation to an orchestra, a drum kit and so on to record ambient sounds and instrumental mixes.

**Figure 4.30.** *Examples of boom stands*

Articulated arms are widely used in the studio, especially in sound booths, and they can be attached just about anywhere. They are very practical, but you have to pay attention to their sturdiness, the fluidity of their joints and the load they can support. A poor quality arm quickly becomes annoying and unusable. They exist with offsets of varying levels of significance and fixing systems of varying levels of sophistication (often clamp type).

**Figure 4.31.** *Examples of articulating arms*

Similar to the articulated arms, but more used in the radio studio, we find the swan neck-type legs that can twist in all directions. Here, too, we will pay attention to the weight that they can support without becoming damaged.

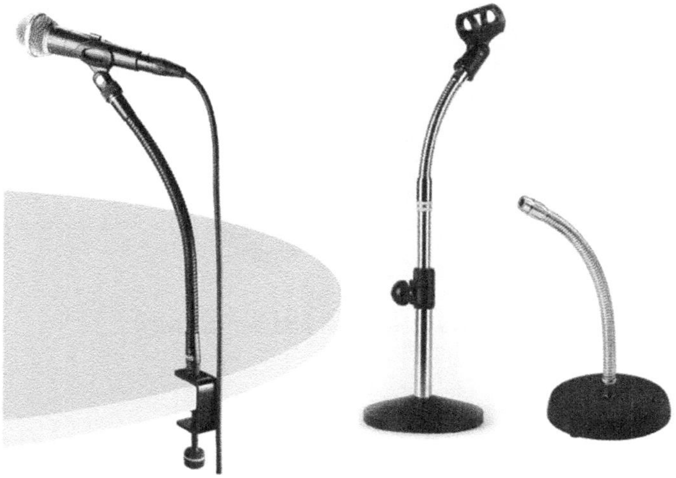

**Figure 4.32.** *Examples of different gooseneck microphones*

The ends of the legs, poles, swan necks, arms and tripods have a thread on which a suspension device can be fixed. There are several possible diameters and threads (3/8'', 5/8'', 1/2'', 1/4''), both on the support and suspension sides. In order to be compatible with each other, there are adapters that can combine all dimensions.

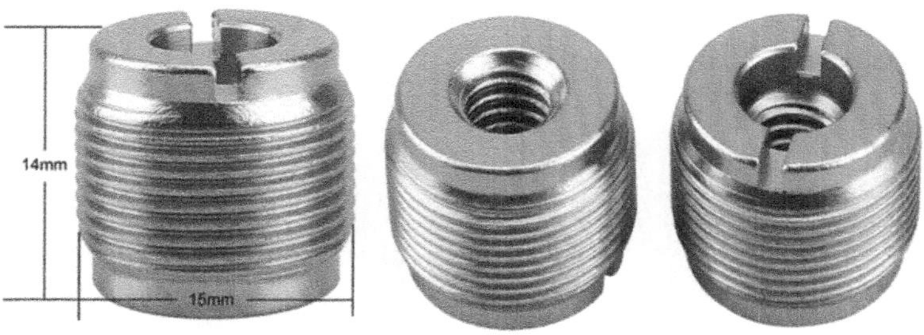


**Figure 4.33.** *Thread adapter for microphone support and suspension*

In some situations, it may be necessary to rotate or tilt the suspension and microphone assembly.

**Figure 4.34.** *Ball joint adapter (Gravity)*

### 4.4.6. *Bonnets*

Every sound recordist has experienced the problem of wind when recording outdoors. It is not the wind itself that produces the unwanted sound, but its interaction with the microphone transducer diaphragm.

To limit this phenomenon, we use a *windshield* or *windscreen*. There are several types of windshields: foam windshields, long-haired windshields, cage windshields and *windjammer windshields*.

Foam windscreens are suitable for the breath of a person speaking and for wind speeds up to 10 km/h. They often come with the purchase of a microphone and cannot be installed on all microphones. Their effectiveness is very relative; they serve more as protection than anything else.

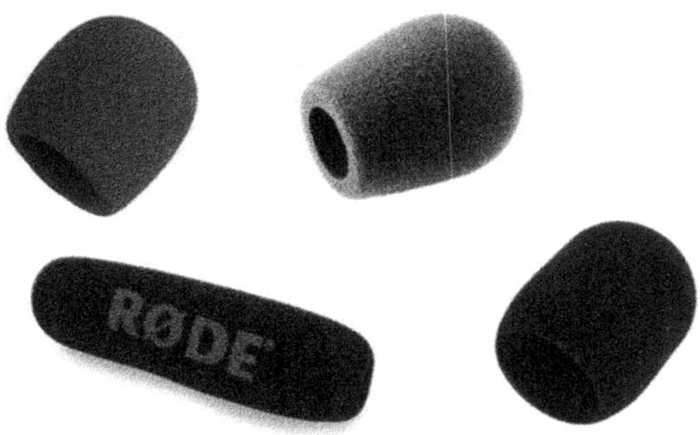


**Figure 4.35.** *Different types of foam cups (Shure, Sennheiser and Rode). For a color version of this figure, see www.iste.co.uk/reveillac/recording.zip*

The long-haired windscreens can withstand winds of up to 30 km/h. They are very useful for reporting and, like the previous ones, they slip over the microphone, which limits their use.

**Figure 4.36.** *Long-hair cap (Rode and Rycote)*

Cages, also known as baskets, are the most effective and also the most expensive. They enclose the entire microphone, which is usually suspended inside. They are available in different diameters and lengths to suit the shape and type of microphone.

**Figure 4.37.** *Cage and windjammer caps (Rycote)*

They have a flexible cable to ensure a connection to the outside and can sometimes have a handle. Usually, they are sold with a windjammer, which is a long-haired fabric sleeve. They are very efficient even with strong winds, and for some models there is even a rain cover.

## 4.5. Conclusion

To make a sound recording in good conditions, the environment is very important. The presence of additional noises, reflections, echoes and reverberations can be disturbing elements. An acoustic treatment is, thus, often necessary.

A word of caution, we are not looking for a deaf (anechoic) room. Far from it – a place filled of a beautiful coloration is often marked with a certain charm. It even characterizes the future recording.

**Figure 4.38.** *An anechoic chamber (CNRS, LMA Laboratory and Aix-Marseille University). For a color version of this figure, see www.iste.co.uk/reveillac/recording.zip*

Remember, it is all in your ear and your ability to listen to the sound. You decide and appreciate the quality of the sound environment.

A good sound treatment is a long-term job. The parameters are complex and numerous; small details can change everything. It is not just a matter of absorbing panels, bass traps, diffusers and accessories, even though you may have chosen the best quality elements. Only your judgment counts. It is up to you to find the right balance.

In the studio and, even more so, live or in a place you do not know, it is your experience that will make the difference.

# Conclusion

The first four chapters that make up Volume 1 of this book have come to an end. I have certainly left out some things that might have seemed useful to you, but above all, I have presented what I know and what my experience has taught me and made me understand over the years in my recording work.

The history of recording is a very vast subject and here I have only summarized the key periods that seemed to be necessary for the description of the temporal evolution of the techniques and equipment for capturing and recording a sound message.

As far as the voice is concerned, I have tried to keep it simple by avoiding drowning the reader in medical and physiological aspects, which are clearly complex and far removed from my knowledge as a sound engineer. The notions approached seem necessary to me in order to have an idea of the vocal instrument and to master the vocabulary attached to it.

The third chapter, dedicated to microphones, is, in my opinion, of the utmost importance. The choice of the recording system is a priority when considering the context, the location, the environment, the live or studio session, the singer, the backing vocals, the voice-over, the musical genre and so on. Just as the photographer chooses the right lens for the shot to be taken, the sound engineer must choose the most appropriate microphone. The parallel does not stop there. A blurred photo will remain of poor quality no matter how it is treated. It is the same for the recording; a bad sound recording will always keep its original defects even though today's techniques can improve it.

When recording a voice, the acoustic environment around the performers is very important. In live situations, we often have to deal with it. On the other hand, in the studio or in certain sound recording situations, we can choose and improve the sound of the place, the room or the premises where the recording will take place. The means to be used are diverse and varied: absorbing panels, bass-traps, diffusers, screens and so on. Associated with these, the accessories of gripping and filtering can contribute to calibrating a vocal sound message in order to obtain the desired coloring. These are the elements that I have dealt with in the last chapter and they are far from being negligible. The voice is a fragile instrument that must be handled and treated with great care and delicacy.

To continue, I invite you to read Volume 2, the content of which is more practical, focusing on studio work, sound processing hardware and software, the configuration of the audio chain for recording and processing, sound recording techniques, sound feedback for the performer, voice mixing and special effects.

# Appendices

# Appendix 1

# Sound Unit

## A1.1. Introduction

The *decibel* (*dB*) is the main unit of measurement in the field of audio, but it is not the only one. It is easily associated with the *pascal*, the *volt* and others, without counting its variations the *dB SPL*, the *dBu*, the *dBv*, the *dBA* and so on.

In the following, you will find some details about all these units and the possible shortcuts to switch from one to another.

## A1.2. Bel and decibel

When the first physicists and inventors became interested in audio phenomena, which gave rise to acoustics and then years later to electro-acoustics, they very quickly needed a way to quantify and measure the perception of a sound signal.

Alexander Graham Bell (1847–1922), an American engineer and scientist, to whom we owe in part[1] the invention of the telephone, was also the person who established the first unit of measurement of an acoustic phenomenon. His mother's deafness and his father's work to teach deaf-mutes (as they were called at the time) to read lips and articulate led him to become fully involved in acoustics.

Bell's approach was based on the ear and its threshold of audibility. After having chosen a reference frequency, 1,000 Hz, he evaluated this threshold for a human being on a rather large population sample.

1 Today, we recognize the anteriority of the invention of the telephone to Antonio Meucci.

His work focused on the minimum and maximum perception levels of a sound signal at this frequency.

On average, for the minimum level perceptible by the ear, the result obtained was close to $10^{-12}$ $W/m^2$ or $10^{-12}$ $W.m^{-2}$ (watt per square meter)[2] or $20$ $\mu Pa$ (micro-pascal[3]) or $2x10$ $N/m^2$ (newton per square meter) and the maximum level acceptable by the ear $10^2$ $W/m^2$ or $10^2$ $W.m^{-2}$ or $200$ $Pa$ (pascal) or $200$ $N/m^2$.

From the difference between the two, Bell established a comparative scale using the lowest audible sound intensity at 1,000 Hz as a reference.

The extreme ratio between minimum and maximum perceived being very large, it is easy to borrow from mathematics the notion of the logarithm, which will allow this scale to be contracted. This is what Bell did by creating the unit that bears his name, the *bel*, which divided by 10, to be more precise, would become the decibel (*dB*).

Taking these minimum and maximum values, we can deduce:

$$Min\ dB\ level = 10\,log\frac{I_0}{I_0} = \frac{10^{-12}}{10^{-12}} = 0db$$

$$Max\ dB\ level = 10\ log\frac{10^2}{10^{-12}} = 140dB$$

with I0, the reference acoustic intensity ($10^{-12}$ $W \times m^{-2}$).

This new scale was graduated between *0* and *140 dB*.

The sound pressure level is defined as 10 times the decimal logarithm of the ratio of the instantaneous sound pressure *P* (in *Pa*), squared, to the reference sound pressure, *Po*, squared (*0.00002 Pa*), that is:

$$10\,log(\frac{P^2}{Po^2})$$

2 The watt per square meter is a unit of the international system of measurement of power flux density.

3 The pascal (Pa) is the unit of pressure in the international system. A pressure of one pascal is a uniform stress acting on a flat surface of 1 square meter, exerting a pressure perpendicular to this surface with a total force of 1 newton.

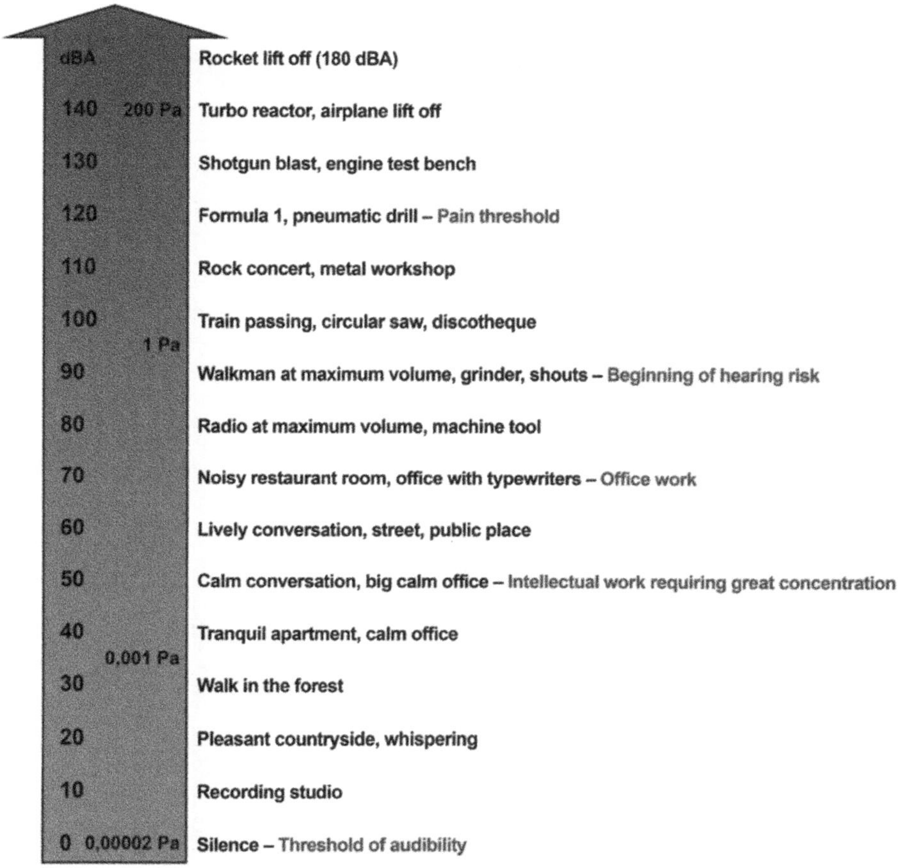


**Figure A1.1.** *Sound level scales (source: Wikipedia). For a color version of this figure, see www.iste.co.uk/reveillac/recording.zip*

The work of the German physician E. Weber[4] and the psycho-acoustician G. Fechner[5] at the end of the 19th century showed that this scale was justified. The Weber–Fechner law states that auditory sensation varies as the logarithm of sound excitation.

4 Ernst Weber, 1795–1878, German physician, precursor of experimental psychology. He worked on the mechanics of wave propagation in the bloodstream and conducted experimental studies on perception.

5 Gustav Fechner, 1801–1887, German philosopher and psychologist known as one of the founders of experimental psychology and psychophysics.

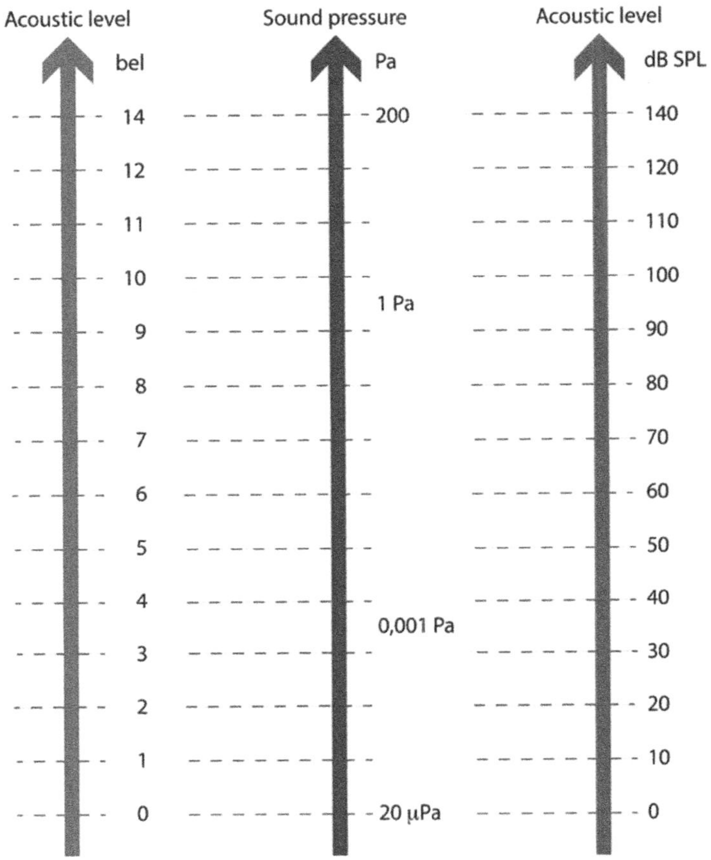


**Figure A1.2.** *Comparison of bel, decibel (dB SPL) and pascal (Pa). For a color version of this figure, see www.iste.co.uk/reveillac/recording.zip*

## A1.3. The different variants of the dB

In Figures A1.1 and A1.2, you may have noticed two new items mentioned *A* and *SPL* following the acronym dB. What do they mean?

When measuring sound levels, the *decibel A* (*dBA*) can be used. Indeed, the human ear has a different sensitivity depending on the frequency of sound signals (low or high). To compensate for this physiological behavior of the ear, the sound powers are weighted according to the frequency components of the sound waves captured before adding them together to provide a global measurement result. The *dB(A)* takes into account this weighting.

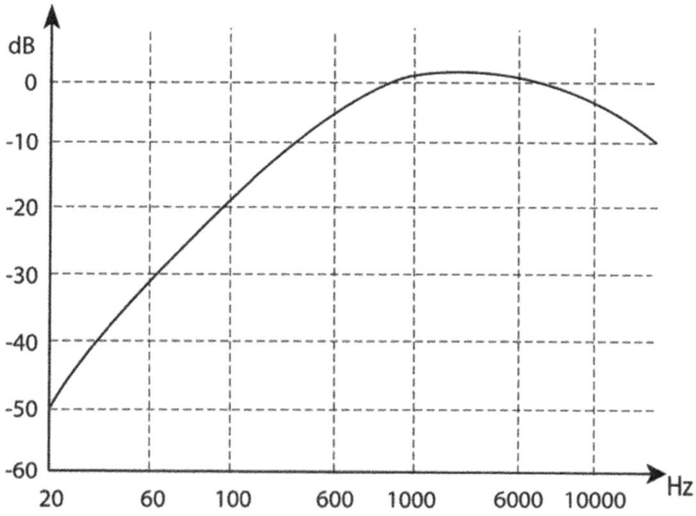


**Figure A1.3.** *Psophometric diagram (weighted curve) dBA*

Compared to *dBA*, the *decibel SPL* (*dB SPL*), which stands for decibel *Sound Pressure Level*, shows a sound level in relation to a reference value that gives the *0 dB* level. This reference, for sound pressure, corresponds to the lowest audible sound, that is, $10^{-12} W/m^2$ or $2 \, x10^{-5}$ *Pa* ($20\mu$ *Pa*) as specified in section A1.2. Two sounds give the same value in decibels *SPL*, if they correspond to the same sound pressure, whether they are low or high pitched. This is the unit that is specified in the characteristics of microphones, since the membrane of a microphone is permanently subject to the influence of pressure when recording a sound signal.

To convert pascals to dB SPL, a calculation that may be needed for some measurements with microphones, the following formula can be applied:

$$dB\ SPL = 20log(P/(2.0x10^{-5}))$$

where *P* is the pressure on the diaphragm in pascals.

To continue, there are many other units such as *dBm*, *dBW*, *dBu*, *dBV*, *dB(B)*, *dB(C)* or *dBFS*.

The *dBm* or *decibel milliwatt* uses a reference value associated with a power of *1 mW* under *600 ohms*.

$$0\ dBm = 1\ mW$$

For the *dbW* or *decibel watt*, we associate a reference value of *1W*.

*0 dBW = 1W = 0dBm + 30 dB = 30 dBm*

To convert a power in decibels, knowing that the reference value *Pref* is *0 dBm = 1mW*, we write:

*Level in dBm = 10 log(P/Pref)*

For example:

*4 mW = 10 log (4/1) = 6 dBm*

*8 W = 10 log (8,000/1) = 39 dBm*

The *dBu* is a unit used to quantify the amplitude of an electrical signal. The dBu measures the voltage in relation to a reference of 0.775 volts RMS[6], which corresponds to the voltage of a 600-ohm load subjected to 1 mW.

For electronics engineers and sound engineers, the sound pressure is the voltage, because the amplifier outputs are in low impedance and the inputs in high impedance; there is, therefore, no adaptation in power. The dBm has no reason to exist; therefore, the dBu is used.

If *U* is the voltage and *U0* (0.775 V) the reference voltage, we have:

*Level in dBu = 20 log U/U0*

Generally, electronic equipment works at a level of *+4dBu*, that is, *1.228V* or *1.78 dBV*.

For the *dBV*, the reference voltage is this time 1 volt RMS (*0 dBV*) or *2.21 dBu*.

*Level in dBV = 20 log(V)*

With regard to the dB(B) and the dB(C), these are variants of the dBA that bring a weighting according to the frequency and also to the intensity.

6 RMS (Root Mean Square) is the square root of the average of this quantity squared, over a given time interval or the square root of the second-order moment of the signal. It is called the RMS value.

The *dBFS* (*dB Full Scale*) designates a maximum amplitude before the clipping of the signal. It is used in the digital domain, and it does not have a universal reference value. It is instead related to the hardware, the manufacturers, and the editors of audio software. The values of *-18* to *-12 dBu* are generally accepted as corresponding to *0 dBFS.*

## A1.4. Conclusion

It is not easy to find one's way around: while A. G. Bell started with an excellent idea, the evolution of technology and science has given rise to many variants of the decibel that are not always easy to understand. What we must remember is that the unit depends on what we are measuring: an acoustic pressure, a power, a voltage, an amplitude and so on. We must always keep in mind that a value in decibels does not designate an absolute value, but a difference in relation to a reference value.

# Appendix 2

# Audio Connectivity

## A2.1. Introduction

Pieces of professional audio equipment are connected to each other by cables that carry different types of signals. Each of them has a more or less standardized but very diverse connector system. This appendix presents most of the plugs, sockets or connectors used by many pieces of equipment, especially microphones.

## A2.2. Coaxial jack

The jack connector is a coaxial connector that has existed for many years (invented in 1877) and is available in different types.

It can be mono (two contacts), stereo (three contacts), multi-pole (four contacts), male, female, chassis according to different diameters:

– ¼''or 6.35 mm;

– 0.206" or 5.23 mm (aviation standard PJ-068);

– 4.4 mm called a "bantam" or "TT";

– 3.5 mm (invented in the 1960s) called "miniature";

– 2.5 mm called "subminiature".

The connection points of a 6.35 mm jack are often designated with the letters "T" (*Tip*), "R" (*Ring*), "S" (*Sleeve*).

A mono jack has two points of contact: the tip and the sleeve, or "TS".

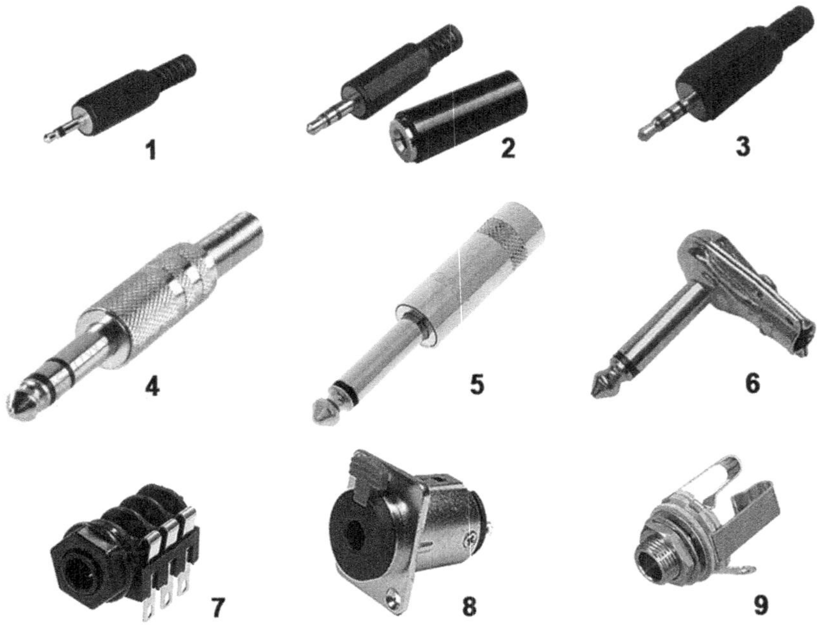

**Figure A2.1**. *Some models of coaxial jacks. 1) subminiature 2.5 mm mono male jack, 2) 3.5 mm miniature stereo male jack, 3) 3.5 mm miniature multi-pole male jack, 4) 6.35 mm (1/4") stereo male jack, 5) 6.35 mm mono male jack, 6) 6.35 mm angled mono male jack, 7) 6.35 mm chassis stereo jack, 8) 6.35 mm chassis mono jack, 9) 6.35 mm chassis mono jack. For a color version of this figure, see www.iste.co.uk/reveillac/recording.zip*

A stereo jack has three points of contact: the tip, the ring and the sleeve, or "TRS", with three possible uses:

– A stereo link with the tip for the left channel, the ring for the right channel and the sleeve for the ground.

– A symmetrical connection (e.g. a microphone) where the tip is the hot point, the ring is the cold point and the sleeve is the ground.

– The insert function of a mixing console with the tip for *send*, the ring for *return*[1] and the sleeve for ground.

A multi-pole jack has four points of contact: the tip, ring 1, ring 2 and the sleeve, or "TRRS".

1 Some manufacturers do not always respect this standard, and so the send and return can be reversed.

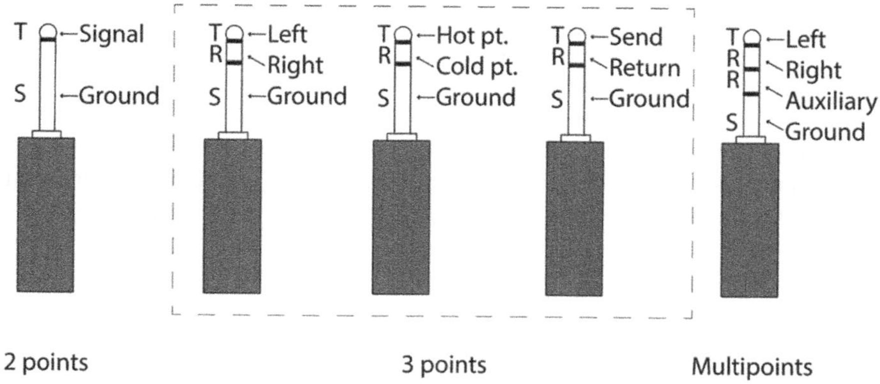


**Figure A2.2.** *The different contact points for the 6.35 mm jack connectors*

Although there are already numerous types of jack, there is another jack of a very particular type, called a "bantam"; it is also called a jack TT (*Tiny Telephone*). The TT connector is frequently used in the audio field. This one has a diameter of 4.4 mm.

In the studio, this connector is often used on professional mixing consoles and *patchbays* because of its compactness and reliability. TT connectors are available in a TRS (*Tip Ring Sleeve*) version for handling balanced signals.

**Figure A2.3.** *A "Bantam" or TT patchbay cable in TRS version*

## A2.3. XLR plug

Its dimensions are standardized via an international specification: IEC61076-2-103. This connector is also called a *cannon* connector from the name of the American company that produced the first series.

The number of pins it has is variable and can range from three to seven pins; however, the three-pin connector is the most used. It is used to transport a balanced monophonic audio signal, which has a hot point, a cold point and a ground.

It has several advantages, including the fact that it does not cause a short-circuit at the connection as would a coaxial connector, and that it is equipped with a safety clip to prevent untimely disconnections. It is robust and exists in both the male and female type.

The AES (*Audio Engineering Society*) standard has defined the following pinout:

– pin 1: ground;

– pin 2: hot point;

– pin 3: cold point.

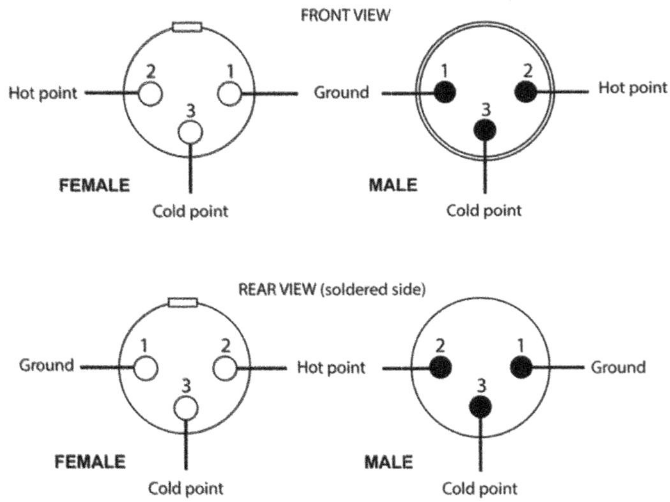


**Figure A2.4.** *Pinout of the 3-pin XLR connectors*

Some older devices may have inverted pins 2 and 3 (especially on American equipment). The numbering is indicated by a marking next to each pin of the socket.

## A2.4. XLR cables and adapter jacks

The most used cable in a studio combines a male XLR plug with a female XLR plug. This cable also has the advantage that it can be used as an extension cable.

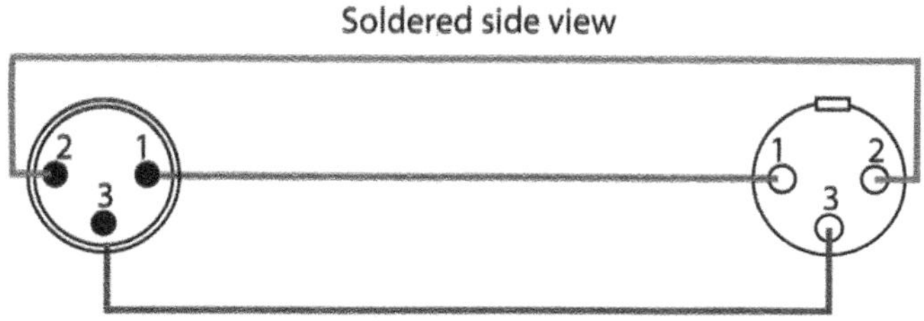


**Figure A2.5.** *Wiring diagram of a classic XLR cable. For a color version of this figure, see www.iste.co.uk/reveillac/recording.zip*

Very often, we have to use both types of connectors – XLR and jack – on the same cable. You will find below the connection diagrams between the two types of connectors, in the balanced and unbalanced modes[2], among others.

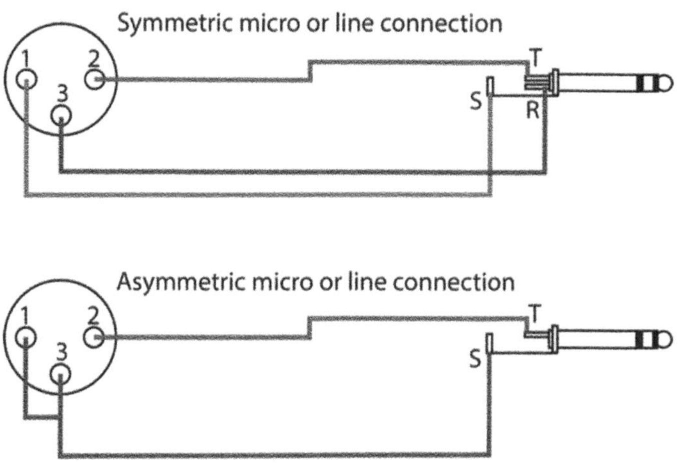


**Figure A2.6.** *Wiring diagram of XLR jack adapters. For a color version of this figure, see www.iste.co.uk/reveillac/recording.zip*

As mentioned in section A2.2, 6.35 mm jacks can be used to create an insert (often audio effects) within a mixing console. This type of cable uses one TRS jack and two TS jacks.

2 A balanced cable has three conductors: a hot point (+), a cold point (-) and a ground. It is protected from interference by the phase opposition between the hot and cold bridges, which cancels out the interfering signals. An asymmetrical cable only has two conductors – a hot point and a ground – and is more subject to interference and parasites.

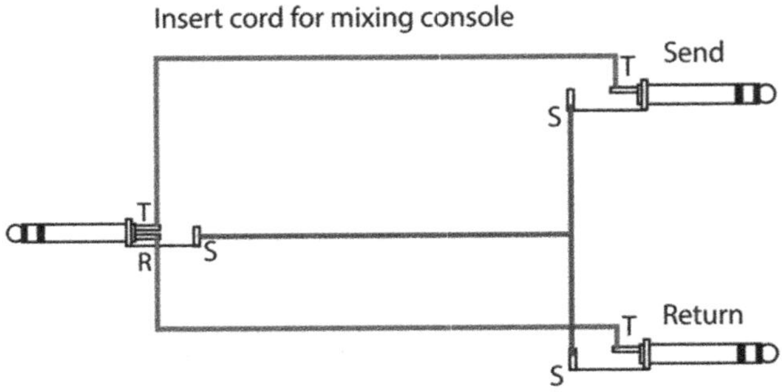


**Figure A2.7.** *Wiring diagram for an insert cord. For a color version of this figure, see www.iste.co.uk/reveillac/recording.zip*

## A2.5. USB connector

A Universal Serial Bus is a connection port for computer peripherals dating from 1996 in its version 1.0.

Over time, versions 1.1, 2.0, 3.0 and 3.1 have appeared, which have mainly improved the throughput:

– USB 1.0: theoretical version that defined the specification;

– USB 1.1 (1998): 1.5 Mbit/s or 12 Mbit/s;

– USB 2.0 (2000): 480 Mbit/s;

– USB 3.0 (2008): 5 Gbit/s;

– USB 3.1 (2013): 10 Gbps.

The connectivity has also evolved over time with different connectors: type A, type B, mini A, mini B, mini AB, micro A, micro B, micro A SuperSpeed (3.0), micro B SuperSpeed (3.0) and type C.

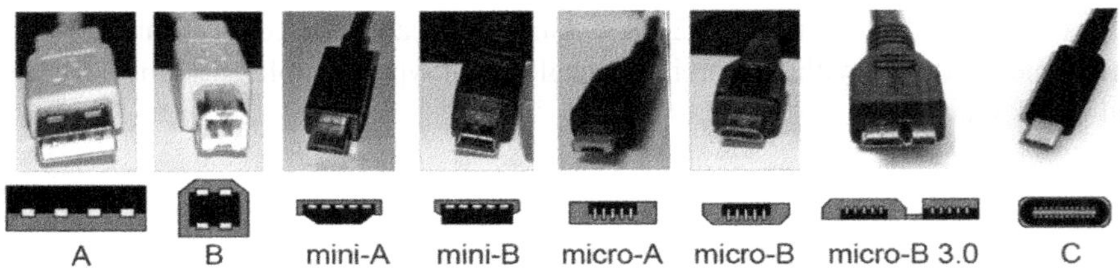


**Figure A2.8.** *Different types of USB connectors*

The latest generation of microphones, called digital, that is, directly connectable to the USB port of a microcomputer or digital equipment (mixer, effects rack, etc.) have become widespread in recent years. The USB port can be used as a power source as well as a communication interface for a digital audio stream.

The pinout varies according to the different connectors. Figure A2.9 shows the A, B, mini-A and micro-B types, which are the most widespread.

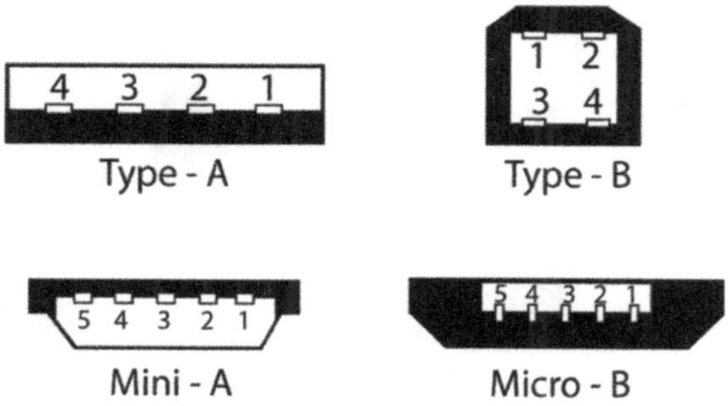


**Figure A2.9.** *Different USB connectors*

| Type-A and type-B connectors | | |
|---|---|---|
| **Pin number** | **Name** | **Function** |
| 1 | VBUS | + 5 volts |
| 2 | D - | Data |
| 3 | D + | Data |
| 4 | GND | Ground |

**Table A2.1.** *USB type-A and type-B pinout*

| Mini-A and micro-B connectors | | |
|---|---|---|
| **Pin number** | **Name** | **Function** |
| 1 | VBUS | + 5 volts |
| 2 | D - | Data |
| 3 | D + | Data |
| 4 | ID | Identification |
| 5 | GND | Ground |

**Table A2.2.** *USB mini-A and micro-B pinout*

## A2.6. Sub-D connector

The Sub-D connector exists with different numbers of pins: 9, 15, 25, 37, 50 and so on. It is referenced with a letter followed by the number of pins: DB25, DA15, DD50 and so on. It has been widely used on computer equipment.

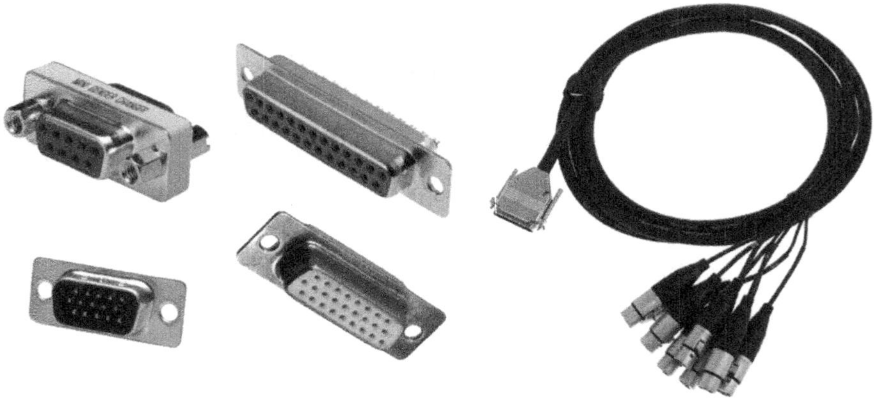

**Figure A2.10.** *Some Sub-D connectors and a female Sub-D/XLR breakout cable. For a color version of this figure, see www.iste.co.uk/reveillac/recording.zip*

On audio equipment, it is also widespread, its compactness making it a connector that takes up little space for a number of important connections. It is often associated with a spreader, which provides multiple connectors of different types including jack, XLR, RCA and so on.

## A2.7. BNC connector

The BNC (Bayonet Neill–Concelman connector) is a connector for coaxial cable. It was designed in the 1940s. Its common uses are analog or digital connections used by various professional equipment, from measurement to antennas, networks, video, audio and so on.

On much studio audio equipment, this connector is used to carry digital signals.

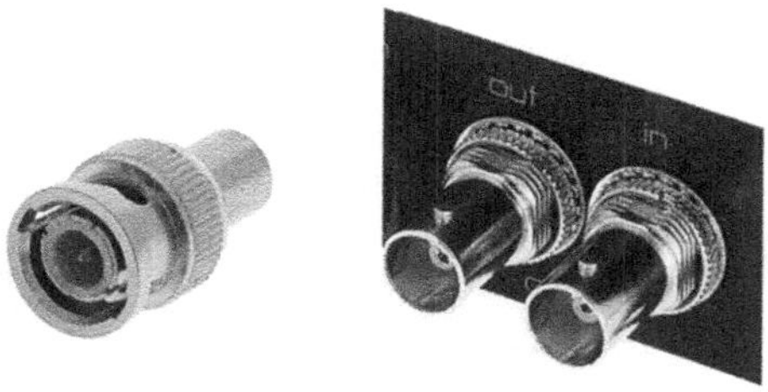


**Figure A2.11.** *BNC male and chassis connectors*

## A2.8. RCA connector

This connector is also called a *Cinch* and is widely used in HiFi and video, where it is generally marked by a color code: white, red, yellow, green, blue, purple, gray, brown and so on.

The S/PDIF standard (Sony Philips Digital InterFace), which is a digital audio data transfer format also uses this connector. The orange color is usually used to identify it.

Some stereo outputs on mixing consoles or instruments also use this connector. The colors red and white are used to define the left and right stereo channels in this order.

**Figure A2.12.** *RCA male and female connectors and chassis. For a color version of this figure, see www.iste.co.uk/reveillac/recording.zip*

## A2.9. Toslink connector

This is an optical connector used in the consumer and professional digital audio field. It was invented in 1993 by Toshiba. The S/PDIF format uses this type of connector when the data transfer is performed by optical fiber. One of its great advantages is its immunity to electromagnetic phenomena.

**Figure A2.13.** *Toslink male and chassis connectors*

There is also a smaller connector called a Mini-Toslink that looks like a 3.5 mm jack.

**Figure A2.14.** *A Toslink/Mini Toslink cable*

## A2.10. Conclusion

The information in this appendix is very simple. Several dozen pages would be needed to describe the audio connectivity in its entirety. For more information, please consult the reference list at the end of this book.

# Appendix 3

# Audio Processing Plugins

## A3.1. Introduction

In the course of this book, I have mentioned the use of plugins many times, especially in Chapter 3 of Volume 2. You will find below a set of tables gathering together the main plugins on the market for each treatment. This content is far from being exhaustive, the list of plugins is long, changes and is often renewed.

If you want to find out more, check the reference list at the end of this book and see what is available from each of the publishers.

## A3.2. Compression plugins

| Publisher | Designation | Remarks |
|---|---|---|
| Audiocation | Compressor AC-1 | Freeware, Compressor<br>Windows: VST |
| Eventide | Omnipressor | Compressor/limiter/expander<br>Windows: VST; RTAS; AAX<br>OSX: AU; VST; RTAS; AAX |
| IK Multimedia | Classic T-RackS Compressor | Compressor<br>Windows: VST; RTAS; AAX<br>OSX: AU; VST; RTAS; AAX |
| MeldaProduction Audio Technologies | MCompressor | Freeware, Compressor/Limiter<br>Windows: VST; VST3; AAX<br>OSX: AU; VST; VST3; AAX |

| Publisher | Designation | Remarks |
|---|---|---|
| PSP Audioware | MasterComp | Compressor<br>Windows: VST; RTAS; AAX<br>OSX: AU; VST; RTAS; AAX |
| Softube | Valley People Dyna-mite | Compressor/limiter/spreader/de-Esser/gate<br>Windows: VST; VST3; RTAS; AAX<br>OSX: AU; VST; VST3; RTAS; AAX |
| URS | 1975 Classic | Compressor/limiter<br>Windows: VST; RTAS; AAX<br>OSX: AU; VST; RTAS |
| Waves | H-Comp | Compressor<br>Windows: VST; VST3; AAX; RTAS<br>OSX: AU; VST; VST3; AAX; RTAS |
| Waves | PuigChild 670 | Compressor, Emulation of Fairchild 660 and 670<br>Windows: VST; VST3; AAX; RTAS<br>OSX: AU; VST; VST3; AAX; RTAS |
| Waves | API 2500 | Compressor<br>Windows: VST; VST3; AAX; RTAS<br>OSX: AU; VST; VST3; AAX; RTAS |
| Waves | SSL G-Master Buss Compressor | Compressor, SSL G-Comp Emulation<br>Windows: VST; VST3; AAX; RTAS<br>OSX: AU; VST; VST3; AAX; RTAS |
| Waves | VComp | Compressor/Limiter/De-Esser, Neve 2254 Emulation<br>Windows: VST; VST3; AAX; RTAS<br>OSX: AU; VST; VST3; AAX; RTAS |
| Waves | CLA-3A | Compressor/Limiter, Universal Audio LA-3A Emulation<br>Windows: VST; VST3; AAX; RTAS<br>OSX: AU; VST; VST3; AAX; RTAS |
| Waves | C1 Compressor | Compressor/expander/gate/equalizer<br>Windows: VST; VST3; AAX; RTAS<br>OSX: AU; VST; VST3; AAX; RTAS |
| Waves | Renaissance Compressor | Compressor<br>Windows: VST; VST3; AAX; RTAS<br>OSX: AU; VST; VST3; AAX; RTAS |

**Table A3.1.** *Some compression plugins*

## A3.3. Equalization plugins

| Publisher | Designation | Remarks |
|---|---|---|
| Abbey Road | TG Mastering Pack | Parametric, four bands + Filters<br>Windows: VST<br>OSX: VST; AU |
| Bomb Factory | Pultec Bundle | Pultec EQP-1A, EQH-2 and MEQ-5 emulation<br>Windows: TDM; RTAS<br>OSX: TDM; RTAS; AU |
| FabFilter | Pro-Q2 | 24 bands + various filters + (Noch filter, tray, etc.)<br>Windows: VST<br>OSX: VST; AU |
| IK Multimedia | T-Racks Classic Equalizer | Parametric, six bands + filters<br>Windows: VST; RTAS; AAX<br>OSX: VST; RTAS; AAX |
| MeldaProduction | MEqualizerLinearPhase | Three equalization algorithms, nine filters<br>Windows: VST; VST3; AAX<br>OSX: VST; VST3; AAX |
| PSP Audioware | Retro Q | Parametric, 3 bands, Vintage EQ Emulation<br>Windows: VST; AAX<br>OSX: VST; AAX; AU |
| RJProjects | Aqualizer | Freeware, 32 Band Graphic, +/- 12dB<br>Windows, OSX: VST |
| Rob Papen | RP-EQ | Eight bands + filters + spectrogram<br>Windows: VST; AAX<br>OSX: VST; AAX; AU |
| Softube | Tube-tech ME1B | Tube-Tech ME-1B emulation based on Pultec MEQ-5 |
| Softube | Tube-Tech PE1C | Tube-Tech PE-1C Emulation |
| Sonnox | Oxford EQ | Parametric, five bands<br>Windows: TDM; RTAS; AAX<br>OSX: TDM; RTAS; AAX |
| SPL | Free Ranger | Freeware, Graphic, four bands<br>Windows: VST; VST3; AAX<br>OSX: VST; VST3; AAX; AU |

| Publisher | Designation | Remarks |
|---|---|---|
| TC Electronic | Assimilator | A curve, Presets, Limiter<br>Windows: VST<br>OSX: VST; MAS |
| Universal Audio | Pultec Passive EQ Collection | Pultec EQP-1A; HLF-3C and MEQ-5 emulation |
| Universal Audio | NEVE 1073 EQ | NEVE 1073 Emulation |
| Voxengo | HarmoniEQ | Parametric, Stereo and 5.1 surround equalizer mastering, Dynamic equalization, Spectrum analyzer<br>Windows: VST; VST3<br>OSX: VST; VST3; AU |
| Voxengo | GlissEQ 3 | Parametric, Stereo and 5.1 surround equalizer, Spectrum analyzer<br>Windows dynamic equalization: VST; VST3<br>OSX: VST; VST3; AU |
| Waves | Linear Phase | Equalizer for mastering, nine types of filters<br>Windows: VST; VST3; AAX<br>OSX: VST; VST3; AAX; AU |
| Waves | Q10 | Paragraphic, ten strips<br>Windows: VST; VST3; AAX<br>OSX: VST; VST3; AAX; AU |
| Waves | H-EQ | Hybrid (parametric + curves), seven types of filters<br>Windows: VST; VST3; AAX<br>OSX: VST; VST3; AAX; AU |
| Waves | RS56 Passive EQ | Passive parametric, three bands, six types of filters<br>Windows: VST; VST3; AAX<br>OSX: VST; VST3; AAX; AU |
| Waves | V-EQ4 | Parametric, four bands, NEVE 1081 console emulation<br>Windows: VST; VST3; AAX<br>OSX: VST; VST3; AAX; AU |

**Table A3.2.** *Some equalization plugins*

## A3.4. De-Esser plugins

| Publisher | Designation | Remarks |
|---|---|---|
| Digitalfishphones | Spitfish | Freeware<br>Windows: VST |
| Eiosis | E$^2$deesser | Windows: VST2; VST3; AAX<br>OSX: VST2; VST3; AU; AAX |
| FabFilter | Pro-DS | Windows: VST; VST3; AAX; RTAS<br>OSX: VST; VST3; AU; AAX; RTAS |
| Nomad Factory | BT Deesser DS-2S | Windows: VST; RTAS<br>OSX: VST; AU; RTAS |
| Sonnox | Oxford SuprEsser | Windows: VST; RTAS<br>OSX: VST; AU; RTAS |
| Waves | De-Esser | Windows: VST; VST3; AAX; RTAS<br>OSX: VST; VST3; AU; AAX; RTAS |
| Waves | Renaissance De-Esser | Windows: VST; VST3; AAX; RTAS<br>OSX: VST; VST3; AU; AAX; RTAS |

**Table A3.3.** *Some De-Esser plugins*

## A3.5. Delay and reverb plugins

| Publisher | Designation | Remarks |
|---|---|---|
| UVI | Sparkverb | Windows, OSX |
| Lexicon | MPX<br>PCM<br>LXP | Windows, OSX<br>Windows, OSX<br>Windows, OSX |
| Eventide | UltraReverb<br>TVerb | Windows, OSX<br>Windows, OSX |
| Universal Audio | EMT 140 | Windows, OSX |
| Valhalla DSP | ValhallaRoom<br>ValhallaPlate | Windows, OSX<br>Windows, OSX |
| Sinusweb | FreeverbToo | freeware, Windows |
| Kresearch | KR-Reverb FSQ | freeware, Windows, OSX |
| Rhythm Lab | Mo'Verb | freeware, Windows |
| TC Electronic | ME30<br>ClassicVerb<br>VSS3<br>DVR2 powercore | Windows, OSX<br>Windows, OSX<br>Windows, OSX<br>Windows, OSX |
| Waves | H-Reverb | Windows, OSX |

**Table A3.4.** *Some reverb plugins*

# Appendix 4

# Tube and JFET Microphone Amplifiers

## A4.1. Introduction

In an active microphone, for which a power supply is needed for the electronics to function, vacuum tubes or JFETs (Junction Field-Effect Transistors) are often used. The purpose of this appendix is to introduce them to you.

## A4.2. Operating principle of tube microphones

Generally speaking, condenser microphones use vacuum tubes. They are there to ensure an impedance conversion.

**Figure A4.1.** *A tube microphone, Neumann U47. The tube can be seen in the center, above the transformer. For a color version of this figure, see www.iste.co.uk/reveillac/recording.zip*

By default, the transducers of condenser microphones provide an audio signal with a very high impedance, which has the major disadvantage of significant degradation during its transport through wires, connectors and even very short cables.

Engineers therefore had the idea of placing triode-type electronic tube behind the transducer to reduce the impedance of the signal.

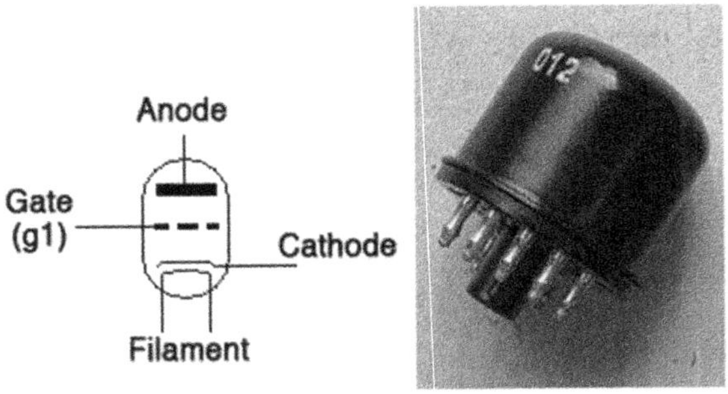


**Figure A4.2.** *Principle of a triode and a Telefunken VF14*[1] *tube designed specifically for microphones*

In a triode tube, there is a filament, a cathode, an anode and a gate. The filament is powered and heats the cathode, which emits negatively charged electrons. These electrons are naturally directed towards the positively charged anode. The flow of electrons thus created has a low impedance.

In a microphone, the gate is considered to be a high-impedance input connected to the transducer. It receives the alternating audio signal and modulates the flow of electrons between the cathode and the anode. To put it simply, if the signal is at its maximum, the flow will be at its maximum; if the signal is minimum, the flow will be minimum.

On this principle, a high impedance signal modulates a low impedance signal, while increasing its amplitude. The tube thus acts as a converter-amplifier of the audio signal.

1 In fact, the VF14 tube is an improved triode, called a pentode, because it has three gates to limit some undesirable effects (secondary emissions from the anode, anode-gate capacitance).

## A4.3. JFET microphones

This type of transistor was invented[2] in 1951 and the first microphone to incorporate this technology was the CMT 20 from Schoeps.

The JFET performs the same function as the triode with a slightly different operation.

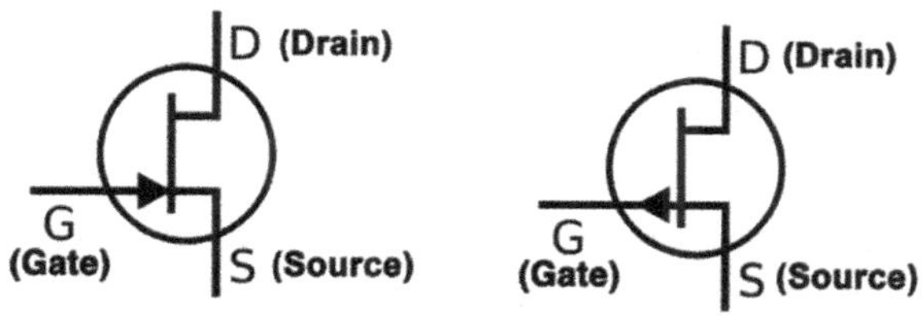


**Figure A4.3.** *Schematic of an FET transistor (M type on the left, P type on the right)*

Within a field effect transistor, there are three pins: the source, the drain and the gate.

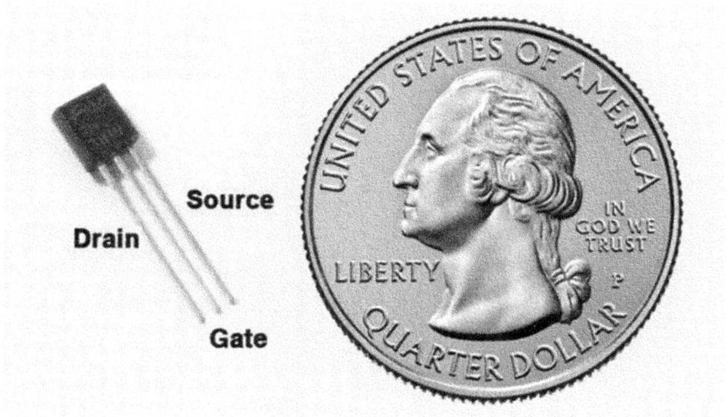


**Figure A4.4.** *A widely used JFET transistor, the 2N3819*

A bias voltage is applied between the source and the drain. The electrons carrying a charge will flow from the source to the drain. The source–drain conductivity is a function of the voltage applied to the gate. This voltage, which comes from the transducer, is alternating since it is the audio signal. The gate is considered as a high impedance input.

2 This invention is credited to William Shockley's team at Bell Laboratory.

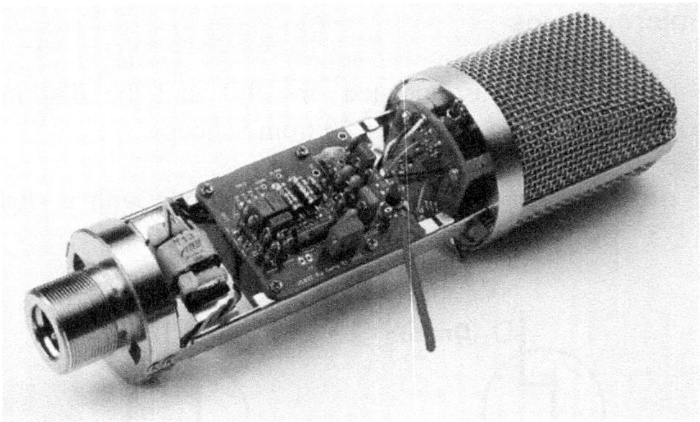

**Figure A4.5.** *A Neumann U87 clone. The arrow indicates the famous JFET (2N3819) mounted on the electronic board. For a color version of this figure, see www.iste.co.uk/reveillac/recording.zip*

The audio signal modulates what circulates between the source and the drain. This link has a low impedance and a higher amplitude transforming the transistor into a converter-amplifier in the same way as the tube presented previously.

# Appendix 5

# Microphone Pairs

## A5.1. Introduction

The use of a pair of microphones makes it possible to record stereophonic sound. There are several techniques, and it is important to know whether the recording you are about to make will be compatible with monophony or not.

The directionality of the microphones used can be omnidirectional, cardioid or hypercardioid, making sure that their directional lobes are homogeneous (no or little alteration due to an obstacle, the microphone itself can be one) and matched.

## A5.2. The AB pair

For this pair, the microphones are separated by 40 cm up to several meters. This is the pair generally used to capture large vocal or orchestral ensembles.

The microphones to be used should be omnidirectional for their excellent bass response and for their directionality, which will highlight the acoustics of the recording location.

If the space is very reverberant or if the distance is very great, cardioid or hypercardioid microphones are preferred.

In an AB pair, the phase of the audio signals is followed and there is no prioritization phenomenon; therefore, the signal is compatible with monophony.

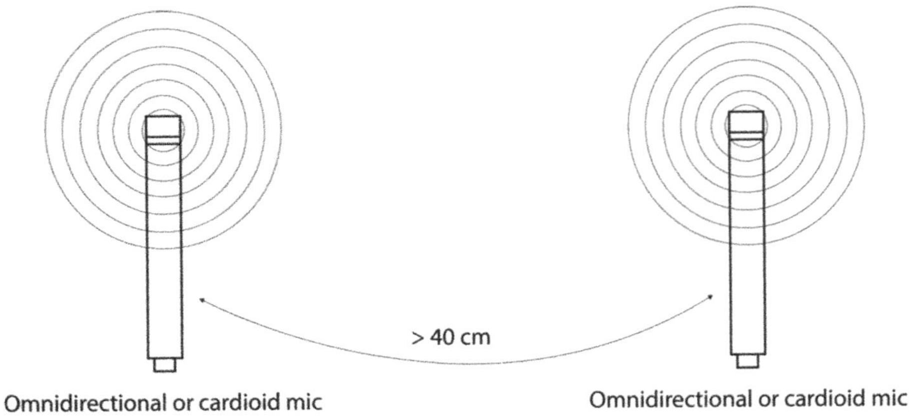


**Figure A5.1.** *Principle of the AB pair*

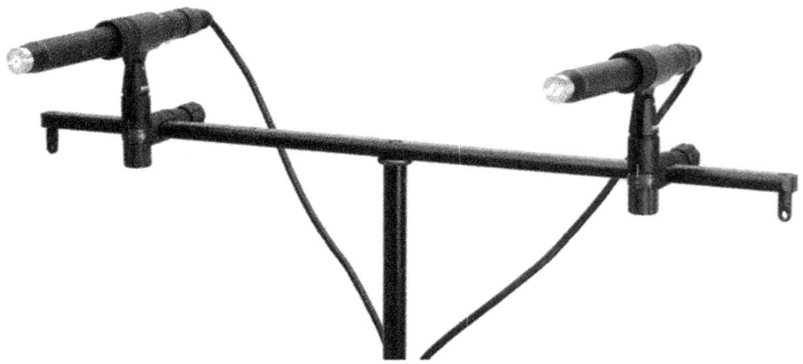

**Figure A5.2.** *An AB pair*

## A5.3. The XY pair

This so-called *phase coincidence* pairing uses two cardioid microphones at a 90° angle with overlapping sensors. This technique is one of the most widely used. It is perfect for close-up pickup and provides a very clear stereo image, with little or no phase problems.

Any differences in the intensity of the audio signal received by the microphones are due to the directionality and orientation of the microphones. The two microphones have perfectly symmetrical directional lobes with respect to the source, so that the received attenuation or boost is identical for each channel.

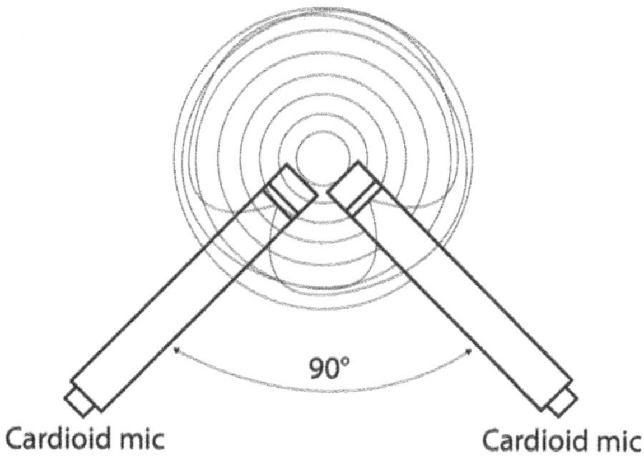


**Figure A5.3.** *Principle of the XY pair*

One of the weakest points is the hole that can exist in the center. Therefore, it is necessary to ensure that the microphones are very homogeneous in their frequency response on each side of their median axis at an angle of at least 45°.

**Figure A5.4.** *An XY pair (source: www.dpa-by-audio2.fr)*

For a source located off the pair's main axis, the signal will not be received in the same area of their directional lobe, so this will generate a disproportion in sensitivity and, thus, in the intensity of the received signal. Since the sensitivity is not the same for all frequencies in the lobe, there is a slight variation in the timbre of the signals received by the microphones, which reinforces the stereophonic effect.

The microphones will be of the cardioid type, generally with a condenser and a small membrane.

The position of the transducers does not introduce any temporal difference, so that compatibility with monophony is possible.

It should be noted that some manufacturers offer XY microphones in a single body (Rode NT4, Superlux E524/D, etc.)

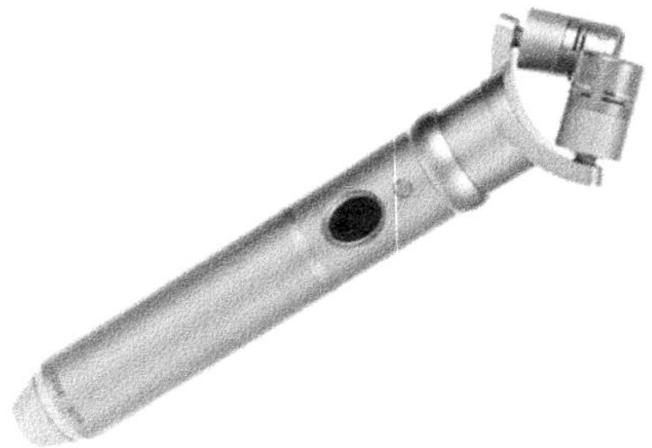

**Figure A5.5.** *The Rode NT4 XY stereo microphone*

## A5.4. The ORTF pair

Unlike the XY pair, the ORTF pair follows the precedence and phase of the signals according to the incidence of the sound source.

By default, the transducers of the microphones are separated by 17 cm, which corresponds to the average distance separating the two ears in humans. This distance can be adapted from a few cm up to 30 cm. The angle formed by the two microphones must be 110° if their directional lobes are homogeneous in terms of frequency response; otherwise, the angle can be varied between 90° and 150°.

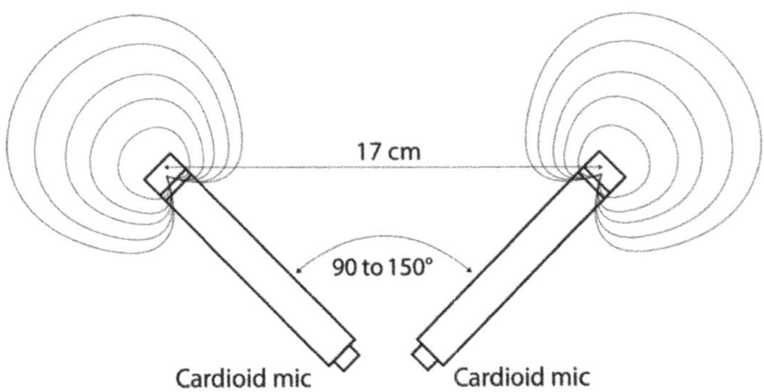


**Figure A5.6.** *Principle of the ORTF pair*

It is important to avoid the hole in the center. For this reason, we choose microphones whose directional lobes are homogeneous over at least 60° on each side of the median axis of the microphones used.

**Figure A5.7.** *An ORTF pair (source: www.harmonycentral.com)*

The DIN pair technique is identical but the angle is 90° and the spacing is 20 cm, as is the NOS (Nederlandse Omroep Stichting, Dutch broadcasting) pair technique with 90° and 30 cm.

## A5.5. The MS pair

MS for *Mid-Side* is the most coherent stereophonic technique in terms of phase. Therefore, the MS couple has a perfect compatibility with monophony. On the other hand, a major problem with this technique is the lack of localization due to phase and precedence. Nothing is ever perfect. Here again it is a compromise.

Another of its strong points is the possibility of modifying the width of the stereophonic field even after the recording is finished, directly during the mixing phase.

The MS setup consists of a cardioid microphone directed at the sound source that will provide the center pan signal and a bidirectional microphone placed at 90° to pick up the side signals.

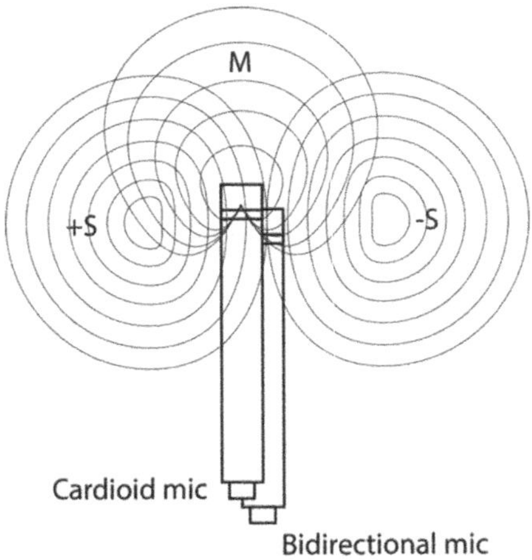


**Figure A5.8.** *Principle of the MS pair*

With this configuration, we get a recording where the amplitude of the two signals determines the width of the stereo space. The signal from the cardioid microphone in the center is placed on one channel and the bidirectional microphone is assigned on the other channel. By varying the ratio between the center and the sides (*Mid* and *Side*), a stereo field of variable width is created. When the amplitude of the central signal increases, the width of the field becomes narrower, while when the sides are stronger, the stereophonic width increases.

**Figure A5.9.** *An MS pair*

With this technique, the sound engineer has more latitude for the final control of the stereophonic image that he or she wishes to integrate into his or her mix, even though the sound recording has been done.

## A5.6. The Decca tree

This technique dates back to the 1950s and was developed by Decca Records sound engineers.

It is mainly used for the recording of large musical formations where atmosphere and spatiality play an important role.

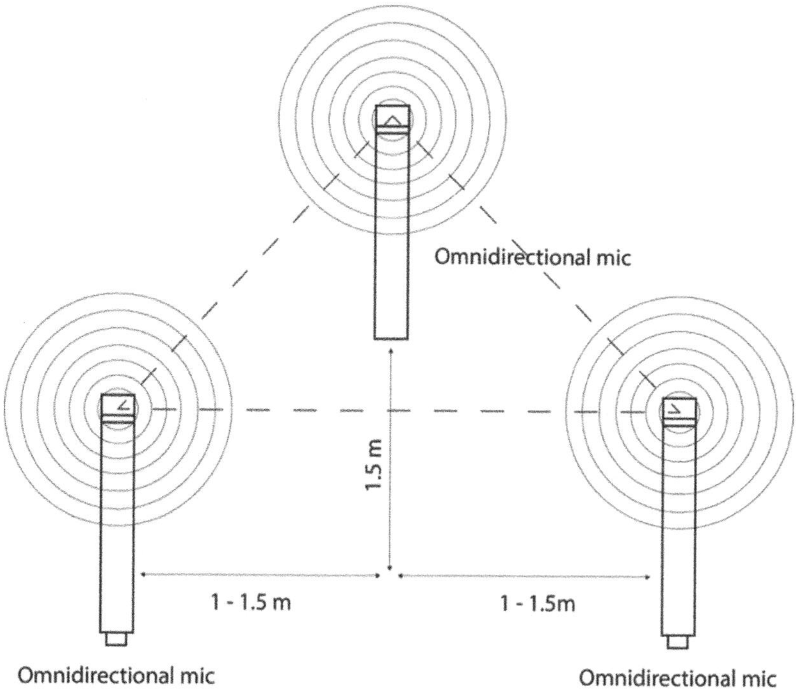


**Figure A5.10.** *Principle of the Decca tree*

The Decca Tree consists of three omnidirectional microphones perched between 2.5 and 3.5 meters high. The central microphone is placed just behind the conductor's head. The side microphones are spaced 2 to 3 meters from the centerline and 1.5 meters from the imaginary line connecting the left and right microphones. The microphones are suspended and tilted downward at an average angle of 30°.

**Figure A5.11.** *Decca tree installation (source: www.musictech.net). For a color version of this figure, see www.iste.co.uk/reveillac/recording.zip*

It should be noted that some people enlarge the installation by placing two other microphones at the extreme right and left, 2 m behind those already existing and spaced between them by 4 to 5 m.

## A5.7. The Blumlein method

This method, named after its creator Alam Blumlein, a British sound engineer, is used for a stereophonic recording that emphasizes the mood of the room.

This pair is very similar to the X-Y technique, but the stereophonic separation is much more pronounced, emphasizing the ambience. It uses two bidirectional microphones which, because of their almost total rejection of lateral fields, capture a coincident stereophonic field.

The captured signal is very wide with a small phase difference, but the rear lobe of each microphone picks up along an opposite axis directed away from the source; this is where the ambience is taken into account.

With a positioning very close to the source, we can obtain a much purer stereophony and this time without any real ambient effect.

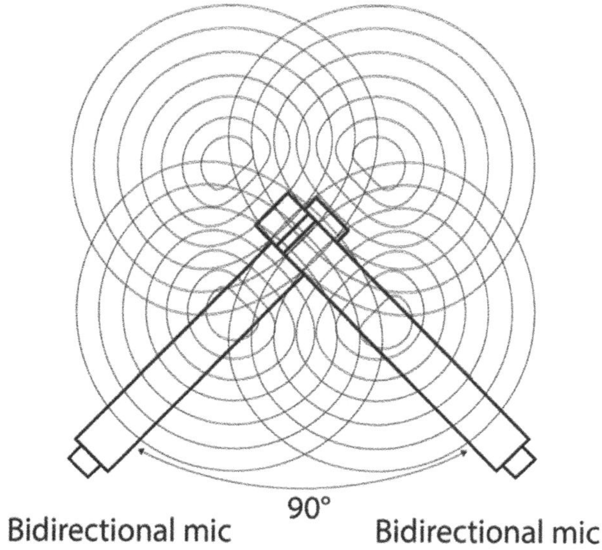


**Figure A5.12.** *Principle of the Blumlein method*

**Figure A5.13.** *A Blumlein installation (source: www.musicradar.com). For a color version of this figure, see www.iste.co.uk/reveillac/recording.zip*

Beware of the walls located towards the back – if they are too close, you will pick up reflections, which can be annoying to various degrees.

## A5.8. The Faulkner Array

This microphone arrangement was invented by Tony Faulkner in the 1970s. It is well suited for recording a small ensemble of musicians in a fairly large room.

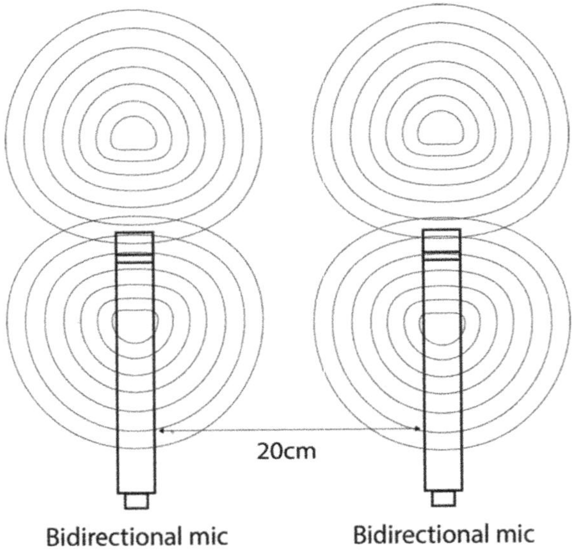


**Figure A5.14.** *Principle of the Faulkner Array*

The pair consists of a pair of bidirectional microphones spaced about 20 cm apart and oriented at an angle of 0°, that is, with their axes parallel.

The two microphones will pick up towards the front and the back taking into account the space of the room and thus the global atmosphere.

## A5.9. The sound-absorbing disk method

This method describes three different techniques: Jecklin Disk, Borud-Shadow and Madson-Shadow.

They all use an isolating screen that separates the two microphones. The sound shadow effect has an influence on the separation of the channels.

The screen is made of a highly absorbent acoustic material.

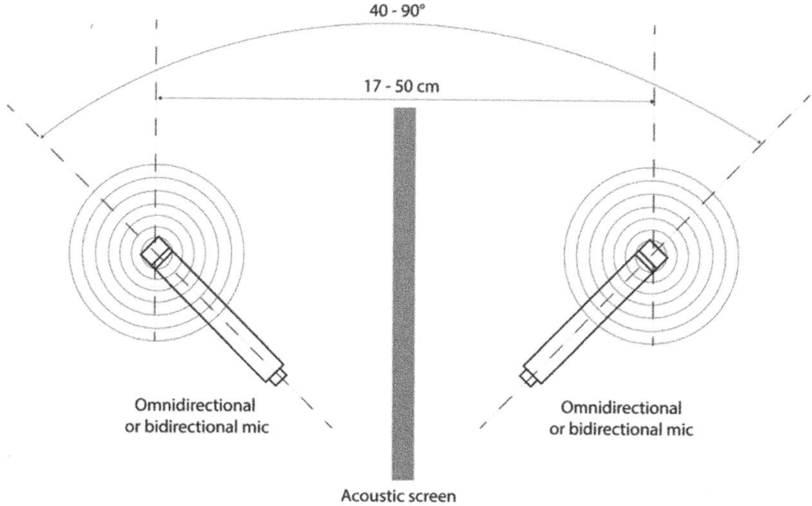


**Figure A5.15.** *Principle of the sound-absorbing disk method*

The Jecklin Disk method, developed by the engineer Jürg Jecklin, is also called OSS (Optimal Stereo Signal). Two omnidirectional microphones are placed 17 cm apart at an angle of 40° and separated by a disk 30 cm in diameter and 2 cm thick.

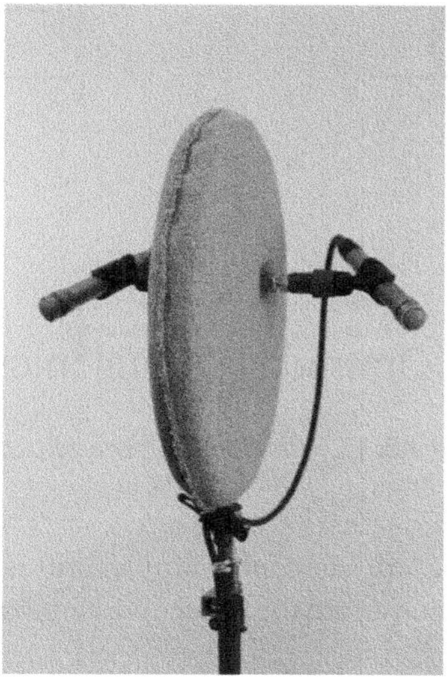

**Figure A5.16.** *A Jecklin Disk installation (source: Wikipedia). For a color version of this figure, see www.iste.co.uk/reveillac/recording.zip*

The Borud-Shadow method is identical to the previous one, but the microphones are separated by 20–25 cm.

The Madson-Shadow method uses two bidirectional microphones at a 90° angle and spaced 30–50 cm apart.

## A5.10. The artificial head

The artificial head is designed to make a so-called binaural sound capture. It consists of using the natural principles of human listening and thus generating a 3D enveloping sound.

Two transducers are placed in the ears of the artificial head. The objective is to capture the sound source with the constraints of diffraction and frequency reflection related to human morphology. With this system, we can analyze and reproduce a three-dimensional space: depth, laterality and elevation.

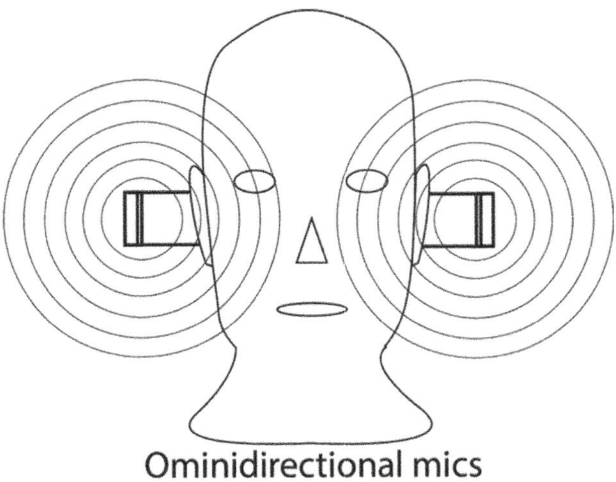


**Figure A5.17.** *Principle of the artificial head*

Listening to this type of stereophonic capture is above all intended for headphones, which will reconstitute the space in a familiar way that is very close to what we hear in reality.

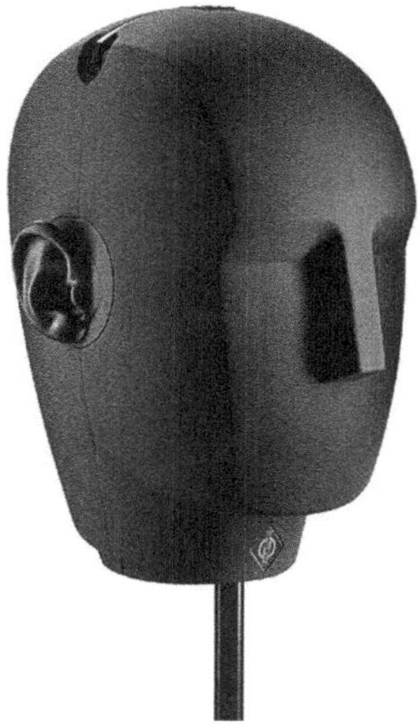

**Figure A5.18.** *Neumann KU100 artificial head*

The system can be so precise that it can induce sonic hallucinations because the rendering is so close to reality.

## A5.11. Conclusion

As far as the choice of microphones is concerned, it is obviously necessary to respect their types (omni, bi, cardioid), after that it is a question of feeling and also of budget; it is here that the ears and intelligence enter into action.

The choice of the right stereophonic pickup pair depends above all on the environment and the experience of the sound recordist; there are always several solutions for the same recording session and it is lucky to call your choice the best.

# Glossary

## A

**Audio Interface:** sometimes called a sound card, this is a computer device that connects microphones, instruments and listening systems to a computer via a USB, PCI, Firewire or Thunderbolt port. It provides analog-to-digital conversion.

**Auto-tune:** tone correction software created by Antares Audio Technologies.

**Auxiliary:** on a mixing console, each of the input channels has one or more sends called "auxiliaries". For these, you can generally switch to "post-fader" or "pre-fader" mode (see later in this Glossary) depending on the type of signal sampling you want to carry out, after or before the fader, the role of which is to manage the output volume of each of the input channels (slices).

## B

**Backing track:** a backing track that makes it possible for a performer to add an additional part to a piece of music. This is particularly the case for a singer. It can be purely rhythmic.

**Baxandall:** audio bass/treble tone corrector invented by Peter Baxandall in 1950.

**Bus:** in the audio field, this is a temporary processing channel, making it possible to gather together different channels or inputs on a mixing console.

**Bus master:** gathers the output channels of a mixing console. Usually, the master bus is copied to the monitoring section for listening in the studio.

**Bypass:** the bypass mode allows an audio signal to pass through a processing device without affecting its sound characteristics.

## C

**Circum-aural:** a headset that completely encloses the ears (see also Intra-aural and Supra-aural).

**Coincident microphones:** a pair of microphones paired to make a sound recording. The microphones are cardioid, of the same type, their sensors (capsules) are as close as possible (less than 40 cm), facing each other, with an angle between 90° and 135° and oriented towards the sound source.

**Compressor:** a hardware or software tool that applies dynamic compression.

– **Dynamic Compression:** an audio technique that reduces the difference in levels between the loudest and softest sounds.

– **Parallel compression:** a fairly simple but very effective technique that consists of taking a track and duplicating the signal to obtain two identical tracks. On the second track, we apply a compression affecting the whole signal, and not just the peaks. This second track is then mixed more or less subtly with the first.

– **Serial Compression:** the technique of using two or more compressors in a light manner to achieve a goal rather than a single compressor acting intensively.

**Crosstalk:** interference between two audio signals.

## D

**DAW (Digital Audio Workstation):** a digital audio workstation (on a computer or tablet) that records, edits, mixes, manipulates, creates, processes and plays back audio content. Avid Pro Tools, MOTU Digital Performer, Magix Samplitude, Cakewalk Sonar, Apple Logic Pro, PreSonus Studio One, and many others are DAWs.

**De-esser:** a tool to reduce the sibilants and occlusives of a sound.

**Decibel (dB):** unit of magnitude, used in acoustics and electronics, defined as 10 times the decimal logarithm of the ratio between two powers.

**Distortion:** undesirable changes to an audio signal.

**DSP (Digital Signal Processing):** this is a microprocessor dedicated and optimized to executing signal processing applications (filtering, extraction, conversion, coding, decoding, compression, etc.).

## E

**Equalizer:** this is a device, piece of equipment or software to correct the timbre of a sound message. It filters, amplifies or reduces certain bands of frequencies that compose the audio signal. It is commonly used for a sound recording, a mixing or a sound system.

## F

**Firewire:** commercial name given by Apple to a multiplexed serial interface, also known as i.Link at Sony, linked to the IEEE 1394 standard.

**FRB:** type of multi-pin connector used in professional audio, especially for sector plugs or patchbays.

## I

**Insert:** this is an access point integrated into the console or mixer, which makes it possible for the user to add and place a new element (peripheral: reverb, compressor, De-Esser, equalizer, etc.) within the audio stream, between the preamplifier and the mix bus.

**Intra-aural:** refers to earphones that fit directly into the ear. Thanks to their small size, they are light and discreet (see also Circum-aural and Supra-aural).

## J

**Jitter:** important parameter of the voice demonstrated by Wendahl, which is characterized by the alteration of the periodicity of the fundamental frequency. A strong alteration brings a hoarse voice (raucity).

## L

**Leaking:** crosstalk between a microphone and the headphones. The microphone picks up sound signals from the headphones.

**Loudness:** numerical quantity that represents the volume of sound perceived by a human being. It is a complex psycho-acoustic concept that takes into account many parameters.

## M

**Mastering:** final stage of audio post-production, which consists of optimizing and calibrating the sound rendering of a mix for its transfer (to a medium such as CD, streaming, for example), its exploitation or its broadcast.

**Matrixing:** another possible name for mastering (see above).

**Melodyne:** audio editor based on musical notes and their characteristics, scales, keys, timing, tempo, sound color, etc. It makes it possible to correct in particular problems related to pitch and accuracy.

**MIDI (Musical Instrument Digital Interface):** protocol, communication standard and file format (MIDIfile) created for electronic musical instruments, officially appeared in 1983. Its connectivity is based on three DIN five-pin connectors: input (MIDI in), output (MIDI out) and redirection (MIDI Thru). Two cables are necessary to ensure bidirectional communication between the different elements of the musical chain.

**Mix:** this is also called audio mixing. In post-production, it is the operation of mixing, within a recording studio, a certain number of sound sources to obtain a coherent complete sound rendering. The mix can also be done live and in the control room for a live event like a concert.

## P

**Pantograph:** an instrument, made of articulated rods, which makes it possible to reproduce a pattern to scale by using the properties of homothety to keep the proportions between the original and the copy.

**Patchbay (audio patch):** a patchbay that centralizes and facilitates access to the signals of most of the components of a studio. It provides an overview of the audio connections rather than having bundles of cables. It can be a 6.35 mm (1/4'') jack or TT, called a bantam or TT (Tiny Telephone), with a diameter of 4.4 mm.

**PCI (Peripheral Component Interconnect):** bus used to connect peripheral cards within a computer from which the PCIe (PCI Express) standard was derived.

**PCM (Pulse Code Modulation):** processing that digitizes an electrical or analog audio signal. After sampling, each sample is quantized and then transformed into a digital code.

**Phoneme:** the smallest distinctive unit making it possible to segment a spoken message. It is a term used in the linguistic field, particularly within phonology.

**Pitch-shifter:** hardware or software tool, part of the sound effects, making it possible to modify the pitch of a sound signal while staying at the same tempo (without modifying its duration).

**Plosive:** a consonant that is pronounced with the mouth closed at the beginning and with an escape of air from the mouth at the end. For example: the letters “p”, “b” and “t” are plosives in English.

**Post-fader:** on a mixing console, the auxiliary “post-fader” send is generally a rotary potentiometer whose role is to take part of the signal after it has passed through the fader (linear potentiometer) of the channel (slice). The send level set on the latter influences the level transmitted to the post-fader potentiometer.

**Pre-fader:** on a mixing console, the “pre-fader” send takes a part of the signal before the fader (linear potentiometer) of the channel (slice). The level of this auxiliary send is not modified in this case if you act on this same fader.

**Presence:** device that adds treble and edge to the sound. Generally, it acts directly on the power stage of the amp via a feedback system (system of reinjection of a part of the signal at the input of the amp).

## Q

**Quadraphony:** this is also called “tetraphony” or “quadrophony”. Its goal is to provide a spatialized sound signal thanks to the use of four independent ways: front-right, front-left, back-right and back-left.

## R

**Raucity:** raucity (harshness or roughness) refers to the sensation and perception of noise in the vocal signal. We speak of the raucity of the voice.

**Re-recording:** technique that consists of recording sound messages by adding them and mixing them with other sequences already recorded.

## S

**Saturation:** for an electronic circuit, saturation occurs when the output level can no longer increase even though the input level increases. The saturation is thus an alteration or deformation of the sound signal that becomes distorted, it generates a distortion.

**Shimmer:** this is a variation in voice amplitude between successive periods when an individual attempts to hold a constant frequency and intensity. Like jitter (see above), it is part of a hoarse voice (see Raucity).

**Sibilance:** whistling due to the presence of strongly accented "sq", "sh", "ch", "z", "j" sounds in speech, also called "sibilants" or "occlusives" of a sound.

**S/PDIF (Sony/Philips Digital Interface):** digital interface that makes it possible for the transfer of digital audio data, also called IEC 958. This standard was designed by Sony and Philips in 1989. It is a direct competitor of the professional standard AES/EBU (Audio Engineering Society/European Broadcasting Union).

**Supra-aural:** refers to headphones that rest directly on the auricle (see also Circum-aural and Intra-aural).

## T

**Talkback:** this is the intercom system used in recording studios to make it possible for the sound engineer or technician in the control room to communicate with people in the recording area or room.

**Thunderbolt**: connection and computer interface format designed by Intel, whose work began in 2007. The Mini Display Port connector, which was already present on Apple computers, was chosen as the standard interface for Thunderbolt. It has a speed of 40 Gbit/s for the latest versions, 3 and 4.

**Timbre:** all the sound parameters that characterize and identify an instrument or a voice.

**Tremolo:** periodic variation of the volume (and/or of the tone) of a sound signal generally produced by a low frequency oscillator (LFO).

**Twang:** lowering of the soft palate or contraction of the pharynx that brings resonance into the nasal cavities. It is also a singing technique to have a stronger voice without effort.

## U

**USB (Universal Serial Bus):** standard defining a serial transmission linking several peripherals to a computer or any other type of compatible device. It was designed in the mid-1990s and replaced many communication ports such as parallel port, serial port and SCSI port. It is linked to a specific connector and has evolved over time by increasing its speed which can theoretically reach 10 Gbits/s, in its latest version, the USB 4 dating from 2017.

## V

**Vacuum tube:** also called "lamp" or "electron tube". Active electronic component generally dedicated to the amplification of a signal. It would be replaced later on by semiconductors (transistor, diode, etc.). It was invented by Lee De Forest in 1907 (triode).

**Vibrato:** periodic variation of the frequency of a sound signal usually produced by a low-frequency oscillator (LFO).

**Voice-over:** the opposite of voice-in. Voice that you hear on a film or video without seeing the person speaking (commentator or narrator, for example).

## X

**XLR:** type of connector, circular section, used to connect professional equipment in the field of audio and lighting. It can have three to seven pins. The three-pin version is by far the most commonly used.

# References


[ADO 13] ADOBE CREATIVE TEAM, *Adobe Audition CC Classroom in a Book*, Adobe Press, 2013.

[BEA 11] BEAUMIER J.L., *L'isolation phonique écologique*, Terre vivante éditions, 2011.

[CAL 19] CALAIS-GERMAIN B., GERMAIN F., *Anatomie pour la voix*, Désiris, 2019.

[CAP 13] CAPLAIN R., *Technique de prise de son*, 6th edition, Dunod, 2013.

[CAS 07] CASE A., *SoundFX: Unlocking the Creative Potential of Recording Studio Effects*, Focal Press, 2007.

[COL 02] COLLECTIF D'AUTEURS, *Le livre des techniques du son*, Dunod, 2002.

[COO 13] COOK F., *Pro Tools 101: An Introduction to Pro Tools 11*, Cengage Learning, 2013.

[COR 19] CORNUT G. *La voix*, Que sais-je ?, 2019.

[COR 20] CORBETT I., *Mic It! Microphones, Microphone Techniques, and their Impact on the Final Mix*, 2nd edition, Routledge, 2020.

[DEC 20] DECKER B., *Template Mixing and Mastering: The Ultimate Guide to Achieving a Professional Sound*, Crowood, 2020.

[ERN 15] ERNOULD F., *Le grand livre du home studio – Tout pour enregistrer et mixer de la musique chez soi*, Dunod, 2015.

[EVE 14] EVEREST A., *Master Handbook of Acoustics*, 6th edition, McGraw-Hill Professional, 2014.

[FOR 86] FORTIER D., *Le mini studio, théorie et pratique*, Fréquences, 1986.

[GER 19] GERMAIN F., GERMAIN B., *Anatomie pour la voix*, Editions Désiris, 2019.

[GIB 97] GIBSON D., *The Art of Mixing*, Mix Books, 1997.

[HUB 13] HUBER D.M., RUNSTEIN R.E., *Modern Recording Techniques*, Focal Press, 2013.

[HUG 05] HUGONNET C., WALDER P., *Théorie et pratique de la prise de son*, Eyrolles, 2005.

[IZH 11] IZHAKI R., *Mixing Audio: Concepts, Practices and Tools*, Focal Press, 2011.

[JAG 18] JAGO M., *Adobe Audition CC Classroom in a Book*, Adobe Press, 2018.

[JOU 99] JOUHANNEAU J., *Notions élémentaires d'acoustique*, Editions Tec & Doc, 1999.

[KAT 14] KATZ B., *Mastering Audio: The Art and the Science*, 3rd edition, Focal Press, 2014.

[KIN 99] KINSLER E.L., FREY R.F., COPPENS A.B., SANDERS J.V., *Fundamentals of Acoustics*, Wiley, 1999.

[LEI 11] LEIPP E., *Acoustique et musique*, Presses de l'Ecole des Mines, 2011.

[MAS 15] MASSY S., JOHNSON C., *Recording Unhinged: Creative and Unconventional Music Recording Techniques*, Hal Leonard Corporation, 2015.

[MAT 69] MATRAS J.J., *L'acoustique appliquée*, Presses Universitaires de France, 1969.

[MER 15] MERCIER D., BOURCET P., CALMET M. *et al.*, *Le livre des techniques du son – Tome 1 : Notions fondamentales*, 5th edition, Dunod, 2015.

[MER 20] MERCIER D., *Logic Pro 10.5 – Apple Pro Training Series: Professional Music Production*, Paechpit Press Publications, 2020.

[MOM 09] MOMMERTZ E., *Acoustics and Sound Insulation*, Birkhauser, 2009.

[NAH 09] NAHMANI D., *Apple Pro Training Series: Logic Pro 9*, Peachpit Press, 2009.

[NEW 12] NEWELL P., *Recording Studio Design*, 3rd edition, Elsevier Focal Press, 2012.

[OWI 13] OWINSKI B., *The Mixing Engineer's Handbook*, Delmar Cengage Learning, 2013.

[PIE 84] PIERCE J.R., *Le son musical*, Belin, 1984.

[PIE 19] PIERCE A.D., *Acoustics: An Introduction to Its Physical Principles and Applications*, 3rd edition, Springer Nature Switzerland AG, 2019.

[REV 18] REVEILLAC J.M., *Musical Sound Effects: Analog and Digital Processing*, ISTE Ltd and John Wiley & Sons, 2018.

[REV 19] REVEILLAC J.M., *Electronic Music Machines: The New Musical Instruments*, ISTE Ltd and John Wiley & Sons, 2019.

[ROG 17] ROGERS J., *Song Mixing Secrets: How to Fix the Most Common Mistakes*, Self-published, 2017.

[ROG 18] ROGERS J., *Audio Mastering Secrets: The Pros Don't Want You to Know*, Self-published, 2018.

[RON 04] RONDELEUX L.J., *Contrôler sa respiration, enrichir son timbre, élargir son registre vocal*, Le Seuil, 2004.

[ROT 02] ROTHERMICH E., *Pro Tools 2020 – The Visual Approach*, Dingding Music, 2002.

[RUM 02] RUMSEY F., MCCORMICK T., *Son et enregistrement*, Eyrolles, 2002.

[SEN 11] SENIOR M., *Mixing Secrets for the Small Studio*, Focal Press, 2011.

[SIE 19] SIEDENBURG K., CHARALAMPOS S., *Timbre: Acoustics, Perception, and Cognition*, Springer Nature Switzerland AG, 2019.

[STR 14] STRONG J., *Home Recording for Musicians for Dummies*, John Wiley & Sons, 5th edition, 2014.

[STR 18] STRONG J., *Pro Tools All-in-One for Dummies*, John Wiley & Sons, 2018.

[TUR 15] TURRIER C., *Le son – Théorie et technologie*, Ellipses, 2015.

[WHI 15a] WHITE P., *L'isolation, la correction acoustique et le monitoring*, Les cahiers de l'ACME, 2015.

[WHI 15b] WHITEMORE T., *Turn Your Spare Room into a Hit Factory*, ACME, 2015.

## Hyperlinks

Hyperlinks are, by essence, volatile. They can change over time, becoming other hyperlinks, even disappearing. All links were checked during the writing of this book but if some no longer work, a quick search on Google or your search engine of choice should help you find them.

### *Software publishers*

Adobe: www.adobe.com

Antares: www.antarestech.com

Audacity: www.audacityteam.org

Avid: www.avid.com

Focusrite: focusrite.com

Fruity Loops: www.image-line.com

Ikmultimedia: www.ikmultimedia.com

Izotope: www.izotope.com

Magix: www.magix.com

MOTU: www.motu.com

Native Instruments: www.native-instruments.com

Nomad Factory: www.nomadfactory.com

PSP Audioware: www.pspaudioware.com

Rob Papen: www.robpapen.com

Softube: www.softube.com

Sonar: www.cakewalk.com

Sonnox: www.sonnox.com

SoundToys: www.soundtoys.com

Spectrasonics: www.spectrasonics.net

Steinberg: www.steinberg.net

TC Electronic: www.tcelectronic.com

Universal Audio: www.uaudio.com

Waves: www.waves.com

## *Hardware manufacturers*

AKG: www.akg.com

Antelope: www.antelopeaudio.com

Apogee: apogeedigital.com

ART: artproaudio.com

Arturia: www.arturia.com

Audient: www.audient.com

Behringer: www.music-group.com/brand/behringer/home

Beyerdynamic: www.beyerdynamic.com

Chandler: chandlerlimited.com

Crane Song Ltd: www.cranesong.com

Crown: www.crownaudio.com

Dangerous Music: www.dangerous.music

DBX: dbxpro.com

DPA: www.dpamicrophones.com

Drawmer: www.drawmer.com

Focusrite: focusrite.com

Genelec: www.genelec.com

IGS Audio: www.igsaudio.com

JBL: www.jbl.com

Klark Teknik: www.klarkteknik.com

Langevin: www.manley,com

Line-6: line6.com

M-Audio: www.m-audio.com

Mackie: mackie.com

Manley: www.manley.com

Midas: www.midasconsoles.com

MOTU: motu.com

Native Instruments: www.native-instruments.com

Neumann: www.neumann.com

Presonus: www.presonus.com

Prism: www.prismsound.com

RME: www.reme-audio.de

Rode: www.rode.com

Roland: www.roland.com

Rupert Neve: rupertneve.com

Sennheiser: www.senheiser.com

Shure: www.shure.com

Soundcraft: www.soundcraft.com

SPL: spl.audio

SSL: www.solidstatelogic.com

Steinberg: www.steinberg.fr

Studer: www.studer.ch

Summit: www.summitaudio.com

T.Bone: www.tbone.mics.com

Tascam: tascam.com

TC Electronic: www.tcelectronic.com

TC Helicon: www.tc-helicon.com

Telefunken: telefunken.com

Tube-tech: www.tube-tech.com

Universal Audio: www.uaudio.com

Warm audio: warmaudio.com

Yamaha: www.yamaha.com

Zoom: www.zoomcorp.com

## Talkback

Microphones and the talkback system: http://mynewmicrophone.com/what-are-talkback-microphones-and-why-are-they-important/

## Plugins

Antares: www.antarestech.com

Celemony: www.celemony.com

Eventide: www.eventideaudio.com

Hornet: www.hornetplugins.com

Izotope: www.izotope.com

Line-6: line6.com

Meldaproduction: www.meldaproduction.com

Sonnox: www.sonnox.som

Universal Audio: www.uaudio.com

Waves: www.waves.com

## Acoustics

Room acoustic treatment and RT60 calculation (in French): http://www.rt60.net/

Distorsion and total harmonic distorsion (in French): http://www.audiosonica.com/fr/cours/post/204/Bruit-Distorsion_et_distorsion_harmonique_totale_THD

Room acoustics (in French): http://physique.unice.fr/sem6/2012-2013/PagesWeb/PT/Reverberation/page1.html

Reverberation time: www.bksv.com/media/doc/BO0228.pdf

Physics – the theory of sound and perception (in French): http://www.sonorisation-spectacle.org/physique/son.html

### Vocoder

Master Class – Vocoders: http://www.emusician.com/how-to/master-class-vocoders

A Brief History of the Vocoder: http://theproaudiofiles.com/history-of-the-vocoder/

The vocoder according to Wendy Carlos: http://www.wendycarlos.com/vocoders.html

### Home Studio

Equipment for a Home Studio: http://www.adventures-in-audio.com/equipping-your-home-recording-studio

Creating your Home Studio (in French): http://www.monter-son-home-studio.fr/

Digital studio and home studio (in French): http://www.studio-musicbox.com/ressources/stud-num-homstud/studio%20numerique%20et%20home%20studio.htm

### Sound effects

History of reverberation (in French): https://www.rts.ch/couleur3/8498599-l-histoire-de-la-reverberation-du-son-.html

What is reverb?: http://www.mediacollege.com/audio/reverb/intro.html

Reverb in audio and music production: http://www.sonible.com/blog/reverb-audio-production/

Echo and reverberation: http://personal.ee.surrey.ac.uk/Personal/P.Jackson/ee1.lab/D3_echo/D3_EchoAndReverberationExpt.pdf

Sound compression (in French): http://blog.landr.com/fr/parametres-de-compression/

What is a compressor for? (in French): http://www.studiocandp.com/blog/post/a-quoi-sert-un-compresseur-audio/20

Dynamic compression (in French): https://www.activstudio.fr/compression-dynamique/

Basics of audio compression (in French): https://www.uaudio.fr/blog/audio-compression-basics/

Beginner's guide – sound compression: http://music.tutsplus.com/tutorials/the-beginners-guide-to-compression--audio-953

EQ or compression?: http://www.sonible.com/blog/first-eq-or-compressor/

Equalization, some benchmarks (in French): http://www.activstudio.fr/eq-equaliseur/

How to use a de-esser: http://www.musicianonamission.com/de-essing-vocals/

De-essing techniques: http://www.soundonsound.com/techniques/techniques-vocal-de-essing

What is de-essing?: http://www.izotope.com/en/learn/the-dos-and-donts-of-de-essing.html

Tips for compression and de-essing:
http://fr.accentsconagua.com/articles/music/compression-tricks-of-the-pros-part-1.html
http://fr.accentsconagua.com/articles/music/compression-tricks-of-the-pros-part-2.html
http://fr.accentsconagua.com/articles/music/compression-tricks-of-the-pros-part-3.html
http://fr.accentsconagua.com/articles/music/compression-tricks-of-the-pros-part-4.html

## *Audio filtering*

Audio filters: http://www2.ece.ohio-state.edu/~anderson/Outreachfiles/AudioEqualizerPresentation.pdf

Audio filters (in French): https://www.sonelec-musique.com/electronique_bases_filtre_audio.html

## *The voice*

Vocal theory – Movement and voice (in French): http://mouvementetvoix.com/theorie.html

Theory of phonation (in French): http://mapage.noos.fr/dixdechoeur/ludivine/Phonation.pdf

Musicality of the spoken voice (in French): http://articles.ircam.fr/textes/Beller05a/index.pdf

Types of phonation:
http://www.ims.uni-stuttgart.de/institut/arbeitsgruppen/ehemalig/ep-dogil/EGG/page10.htm
http://gordon.faculty.linguistics.ucsb.edu/phonation.pdf

## *Audio software tutorials*

Adobe Audition tutorial: http://helpx.adobe.com/en/audition/tutorials.html

Avid Pro Tools tutorial (in French): http://www.cours-gratuit.com/cours-pro-tools/

Pro Tools online tutorials:
http://www.pluralsight.com
http://www.udemy.com
http://www.pro-tools-expert.com/home-page/2017/12/5/tutorial-getting-started-with-pro-tools

Pro Tools reference guide:
http://compositiondemao.com/wp-content/uploads/2015/12/PT_11_RG_FR.pdf
http://resources.avid.com/SupportFiles/PT/Pro_Tools_Reference_Guide_2018.12.pdf

Pro Tools from A to Z: http://curdt.home.hdm-stuttgart.de/PDF/PTExpert_Julian%20Rodgers_The%20A-Z%20Of%20Pro%20Tools.pdf

Logic Pro X online tutorial:
https://fr.tuto.com
https://www.udemy.com/

How to mix a voice with Logic (in French): http://www.zikinf.com/forums/comment-mixer-une-voix-logic-pro-t5151211.html

Elephor: http://www.elephorm.com/audio-mao.html

MJ tutoriels: http://www.mjtutoriels.com/18-techniques-audio

Tutorom: http://www.tutorom.fr/categories-de-tutoriels/fr/audio

Virtual product school (VPS) (in French): http://www.tutoriels-mao.com/les-tutoriels/mix-et-master-de-a-%C3%A0-z-avec-des-plugins-gratuits-detail

Learning mixing with Ableton (in French): http://le-son-ableton.fr/apprendre-le-mixage-avec-ableton/

Les tutos d'Anto: http://www.tutodanto.com/c/voir-tous-les-tutos-mao

Tuto.com: http://fr.tuto.com/tuto/audio-mao/

Home studio débutant: http://www.home-studio-debutant.com/

Joshua Casper tutorial: http://www.joshuacasper.com/

### *Interfaces and communication*

Digital interfaces: http://www.soundonsound.com/techniques/digital-interfacing

Digital audio connections: http://www.mamosa.org/jenfi.home/debuter/connexionsnumeriques.php

### *General sites*

Recording and mixing: http://mixmag.net

Musical materials and more: http://www.pmtonline.co.uk

Musical materials: http://www.musicradar.com

MusicTech, a site dedicated to sound and musical production engineers: http://www.musictech.net/

ProSound: http://www.prosoundnetwork.com/

Audio recording forum: http://homerecording.com

SoundClick, a site dedicated to musical audio: http://www.soundclick.com

Tape Op magazine: http://tapeop.com/

Activstudio: http://www.activstudio.fr/mixage-audio/

Mastering addict: http://mastering-addict.com/#sthash.yun06RU3.dpbs

Training and resources for sound lovers: http://deveniringeson.com/

Zikinf, a general site for music: http://www.zikinf.com/

Thomann, an online music products seller: http://www.thomann.de/gb/index.html

Woodbrass, an online music products seller: http://www.woodbrass.com/

Canford, an audio and video product seller (in French): http://www.canford.fr/

Tape Op magazine: http://tapeop.com/

Keyboard magazine: http://www.keyboardmag.com/

KR Home studio, the magazine for musical creation (in French): http://www.kr-omestudio.fr/

Pro Audio Review magazine: http://www.prosoundnetwork.com/article.aspx?articleid=39995

ProSound website: http://www.prosoundnetwork.com/

# Index

**E, F**



**G, H**

## P, Q



## R, S

Other titles from

iSTE

in

Waves

## 2021

DAHOO Pierre-Richard, LAKHLIFI Azzedine
*Infrared Spectroscopy of Symmetric and Spherical Top Molecules for Space Observation 1 (Infared Spectroscopy Set – Volume 3)*

*Infrared Spectroscopy of Symmetric and Spherical Top Molecules for Space Observation 2 (Infrared Spectroscopy Set – Volume 4)*

## 2020

DANIELE Vito G., LOMBARDI Guido
*Scattering and Diffraction by Wedges 1: The Wiener-Hopf Solution - Advanced Applications (Waves and Scattering Set – Volume 1)*

*Scattering and Diffraction by Wedges 2: The Wiener-Hopf Solution - Advanced Applications (Waves and Scattering Set – Volume 2)*

SAKHO Ibrahima
*Introduction to Quantum Mechanics 2: Wave-Corpuscle, Quantization & Schrödinger's Equation*

## 2019

BERTRAND Pierre, DEL SARTO Daniele, GHIZZO Alain
*The Vlasov Equation 1: History and General Properties*

DAHOO Pierre-Richard, LAKHLIFI Azzedine
*Infrared Spectroscopy of Triatomics for Space Observation*
*(Infrared Spectroscopy Set – Volume 2)*

RÉVEILLAC Jean-Michel
*Electronic Music Machines: The New Musical Instruments*

ROMERO-GARCIA Vicente, HLADKY-HENNION Anne-Christine
*Fundamentals and Applications of Acoustic Metamaterials: From Seismic to Radio Frequency*
*(Metamaterials Applied to Waves Set – Volume 1)*

SAKHO Ibrahima
*Introduction to Quantum Mechanics 1: Thermal Radiation and Experimental Facts Regarding the Quantization of Matter*

## 2018

SAKHO Ibrahima
*Screening Constant by Unit Nuclear Charge Method: Description and Application to the Photoionization of Atomic Systems*

## 2017

DAHOO Pierre-Richard, LAKHLIFI Azzedine
*Infrared Spectroscopy of Diatomics for Space Observation*
*(Infrared Spectroscopy Set – Volume 1)*

PARET Dominique, HUON Jean-Paul
*Secure Connected Objects*

PARET Dominque, SIBONY Serge
*Musical Techniques: Frequencies and Harmony*

RÉVEILLAC Jean-Michel
*Analog and Digital Sound Processing*

STAEBLER Patrick
*Human Exposure to Electromagnetic Fields*

## 2016

ANSELMET Fabien, MATTEI Pierre-Olivier
*Acoustics, Aeroacoustics and Vibrations*

BAUDRAND Henri, TITAOUINE Mohammed, RAVEU Nathalie
*The Wave Concept in Electromagnetism and Circuits: Theory and Applications*

PARET Dominique
*Antennas Designs for NFC Devices*

PARET Dominique
*Design Constraints for NFC Devices*

WIART Joe
*Radio-Frequency Human Exposure Assessment*

## 2015

PICART Pascal
*New Techniques in Digital Holography*

## 2014

APPRIOU Alain
*Uncertainty Theories and Multisensor Data Fusion*

JARRY Pierre, BENEAT Jacques N.
*RF and Microwave Electromagnetism*

LAHEURTE Jean-Marc
*UHF RFID Technologies for Identification and Traceability*

SAVAUX Vincent, LOUËT Yves
*MMSE-based Algorithm for Joint Signal Detection, Channel and Noise Variance Estimation for OFDM Systems*

THOMAS Jean-Hugh, YAAKOUBI Nourdin
*New Sensors and Processing Chain*

TING Michael
*Molecular Imaging in Nano MRI*

VALIÈRE Jean-Christophe
*Acoustic Particle Velocity Measurements using Laser: Principles, Signal Processing and Applications*

VANBÉSIEN Olivier, CENTENO Emmanuel
*Dispersion Engineering for Integrated Nanophotonics*

**2013**

BENMAMMAR Badr, AMRAOUI Asma
*Radio Resource Allocation and Dynamic Spectrum Access*

BOURLIER Christophe, PINEL Nicolas, KUBICKÉ Gildas
*Method of Moments for 2D Scattering Problems: Basic Concepts and Applications*

GOURE Jean-Pierre
*Optics in Instruments: Applications in Biology and Medicine*

LAZAROV Andon, KOSTADINOV Todor Pavlov
*Bistatic SAR/GISAR/FISAR Theory Algorithms and Program Implementation*

LHEURETTE Eric
*Metamaterials and Wave Control*

PINEL Nicolas, BOURLIER Christophe
*Electromagnetic Wave Scattering from Random Rough Surfaces: Asymptotic Models*

SHINOHARA Naoki
*Wireless Power Transfer via Radiowaves*

TERRE Michel, PISCHELLA Mylène, VIVIER Emmanuelle
*Wireless Telecommunication Systems*

## 2012

LALAUZE René
*Chemical Sensors and Biosensors*

LE MENN Marc
*Instrumentation and Metrology in Oceanography*

LI Jun-chang, PICART Pascal
*Digital Holography*

## 2011

BECHERRAWY Tamer
*Mechanical and Electromagnetic Vibrations and Waves*

BESNIER Philippe, DÉMOULIN Bernard
*Electromagnetic Reverberation Chambers*

GOURE Jean-Pierre
*Optics in Instruments*

GROUS Ammar
*Applied Metrology for Manufacturing Engineering*

LE CHEVALIER François, LESSELIER Dominique, STARAJ Robert
*Non-standard Antennas*

## 2010

BEGAUD Xavier
*Ultra Wide Band Antennas*

MARAGE Jean-Paul, MORI Yvon
*Sonar and Underwater Acoustics*

## 2009

BOUDRIOUA Azzedine
*Photonic Waveguides*

BRUNEAU Michel, POTEL Catherine
*Materials and Acoustics Handbook*

DE FORNEL Frédérique, FAVENNEC Pierre-Noël
*Measurements using Optic and RF Waves*

FRENCH COLLEGE OF METROLOGY
*Transverse Disciplines in Metrology*

## 2008

FILIPPI Paul J.T.
*Vibrations and Acoustic Radiation of Thin Structures*

LALAUZE René
*Physical Chemistry of Solid-Gas Interfaces*

## 2007

KUNDU Tribikram
*Advanced Ultrasonic Methods for Material and Structure Inspection*

PLACKO Dominique
*Fundamentals of Instrumentation and Measurement*

RIPKA Pavel, TIPEK Alois
*Modern Sensors Handbook*

## 2006

BALAGEAS Daniel *et al.*
*Structural Health Monitoring*

BOUCHET Olivier *et al.*
*Free-Space Optics*

BRUNEAU Michel, SCELO Thomas
*Fundamentals of Acoustics*

FRENCH COLLEGE OF METROLOGY
*Metrology in Industry*

GUILLAUME Philippe
*Music and Acoustics*

GUYADER Jean-Louis
*Vibration in Continuous Media*